Essential Architecture and Principles of Systems Engineering

Essential Architecture and Principles of Systems Engineering

C.E. Dickerson
Siyuan Ji

CRC Press
Taylor & Francis Group
Boca Raton London New York

CRC Press is an imprint of the
Taylor & Francis Group, an **informa** business

First Edition published 2022
by CRC Press
6000 Broken Sound Parkway NW, Suite 300, Boca Raton, FL 33487-2742

and by CRC Press
2 Park Square, Milton Park, Abingdon, Oxon, OX14 4RN

Library of Congress Cataloging-in-Publication Data
A catalog record has been requested for this book

ISBN: 978-1-138-55017-9 (hbk)
ISBN: 978-1-032-10096-8 (pbk)
ISBN: 978-1-003-21363-5 (ebk)

DOI: 10.1201/9781003213635

Typeset in Times
by KnowledgeWorks Global Ltd.

This book is dedicated to

*The memory of Dr William Etheridge and Professor Peter Smith,
without whom this contribution to architecture and systems would
not have been possible; and to my brother, Dennis Dickerson,
for his enduring encouragement through the entire journey.*

– CD

*Fengling Jiang who inspired me to study physics but never had
the chance to ask, "what is Architecture"; and to my beloved
Jingjing Cao for cooking the most delicious food ever.*

– SJ

Contents

PART I   Architecture and Systems in the 21st Century

PART II   *Tutorial Case Study: End-to-End Demonstration*

PART III   Modelling Languages

PART IV   Case Studies for Practice and Practitioners

Preface

An expansive advancement of systems engineering tools and languages over the past decade has been fuelled by defence, aerospace, and other systems developers working with standards organisations and suppliers of commercial tools for modelling architecture and systems. However, for the non-specialist, and even for practicing engineers across traditional disciplines ranging from mechanical and electrical engineering to manufacturing, the subject of systems engineering remains cloaked in jargon and a sense of mystery.

This need not be the case for any reader of this book and for students no matter what their background is. The fundamental concepts and principles of architecture and systems engineering are simple and intuitive. They can be understood by anyone who will take time to reflect upon them. Systems are distinguished from traditional viewpoints on products and services by a focus on wholeness and interrelationships between parts. System architecting is concerned with structuring the system and its interrelationships in the context of its environment in ways that enable functionality. Modelling is the interpretation of concepts into structures that represent the system. Engineering is concept realisation that uses the models of science to solve a problem or exploit an opportunity in a way that delivers utility and market value.

Essential Architecture and Principles of Systems Engineering shows how systems architecting, modelling, and engineering can be bound into a straight-forward process for architecture and system definition of solutions to complex problems that are implemented by rigorous methods. This fulfils a need for an easily accessible yet rigorous end-to-end approach to exploit the advances that have been made in tools and languages. This goes beyond traditional systems engineering methods. The book embodies a decade of research and very successful academic instruction to students that includes practicing engineers from a variety of industries and government functions. The government-sponsored research with commercial partners crosses multiple industry and engineering boundaries. The material has been continuously improved and evolved from its basis in defence and aerospace towards the engineering of commercial systems with an emphasis on speed and efficiency. The book has also been structured for remote delivery; the material has been successfully delivered online.

The result is a state-of-the-art introduction to the concepts and practice of architecture and systems engineering that has been very successful at both the advanced undergraduate and introductory postgraduate levels. The material is organised for dual use, both as a traditional lecture series and also for remote delivery. The previous volume, *Architecture and Principles of Systems Engineering*, offered a practical engineering perspective that was reflected in several case studies of real-world systems. A decade of feedback from academic and industry users of the previous volume overwhelmingly indicated that it was unique and invaluable for its practical case studies that embodied and demonstrated the concepts and principles of architecture and systems engineering.

The same practical perspective is echoed in this volume. Two new case studies for student practice have been developed (Part IV of the book). One uses an urban traffic management system that is information intensive. It is based on a commercial case study. The other uses an actuator system of systems that is more mechanical in nature. Each of the case studies is formed of two or three practical exercises that students can do in small groups in just a few hours. The student case studies are complemented by a tutorial case study on an air traffic control system that demonstrates the key concepts. The book is self-contained. Students with no background in the case studies or in systems engineering are surprised by how quickly and systematically they can enter a new problem domain and then deliver useful system solutions.

As with the previous volume that was written over a decade ago when the community was rethinking systems engineering, this volume begins with an explanation of the foundations of architecture and systems. The first two chapters offer a broad understanding of the logical and scientific basis of systems engineering in the context of its modern history. The concepts and language are made concrete using real-world systems. This will let you, the reader, speak knowledgably with the experts no matter what your background is.

Part I then proceeds in bite-sized steps to introduce a model-based process and methods for architecture and system definition that are aligned to international standards for the engineering of modern products and services. This part is suitable for a comprehensive short course on model-based systems engineering that is accessible to a general audience. An end-to-end demonstration of the process and methods is provided in Part II using the tutorial case study, which can be understood on its own. The reader can easily progress from the foundational material in Part I to the tutorial in Part II.

Part III provides a concise treatment of modern internationally adopted modelling languages for software and systems: the object-oriented Unified and Systems Modeling Languages (UML and SysML). The relation of these languages to the processes and methods in Part I is also explained. The first three sections of the book, therefore, unfold chapter by chapter with increasing levels of precision and rigor for using UML and SysML that are suitable for a commercial practice of software and systems engineering in a holistic way. This organisation of the material in the book forms a comprehensive core that practising engineers can use as a desktop reference.

Essential Architecture and Principles of Systems Engineering is a book for everyone interested in systems and the modern practice of engineering.

Acknowledgements

The material in this course of study is from a collaborative effort by Professor C.E. Dickerson of Loughborough University and Dr Siyuan Ji at the University of York. We each wish to express special thanks to the postgraduate students whose collaborations have contributed to the advances in this volume of *Essential Architecture and Principles of Systems Engineering*. The collaborative research with Stanislav Rudakov in his postgraduate studies on safety-critical commercial nuclear standards provided an early validation of relational orientated methods for architecture and systems, as did the collaborative research with Zahir Ismail on a formal critical analysis of international standards on systems engineering. His consultancy company became the Platform Sponsor for the UML Profile for ROSETTA (UPR) with the Object Management Group. This was adopted as an international standard in 2018. Most recently amongst the postgraduate students is Rosmira Roslan for her collaborative research on integrating UML with the propositional calculus of logic for the reliability of safety-critical systems. All these contributions have been published in international journals and sources.

Several Research Associates at Loughborough University have also contributed to the delivery of this course of study and made the case studies both understandable and popular with the students: Hao Liang, Jonathan Wilson, Ce Liang, and Jack Day. Each has been involved with applying concepts of the case studies in the book to the commercialisation of wireless charging for electric vehicles. They have provided invaluable feedback on best practices based on this experience; especially Jack, who saw the material first as a postgraduate student and then as a Research Associate. He and Jonathan have both proofread the entire manuscript from this unique vantage point. Jack has since transitioned into a systems engineering position in aerospace. Jonathan has joined the faculty at Loughborough University to lecture in Systems and Mechanical Engineering. Hao has taken a senior systems engineering position with a leading automotive business in Beijing.

The research contributions upon which this book is based would not have been possible without support from the following organisations: the Royal Academy of Engineering, the Engineering and Physical Sciences Research Council, BAE Systems, Jaguar Land Rover, the government office Innovate UK, and the Office for Zero Emissions Vehicles.

An overview of the grants, contributions, and research strategy that underpins the book can be found in Annex A-4. There is a breadth of other topics that complement the book, such as architecture frameworks, styles, patterns, dynamic simulations, and development approaches (spiral, bottom-up, etc.), that were not included in order to keep the book as concise as possible both in length and content. These and other materials to include lecture slides for educators are provided on the Taylor & Francis resource website for the book.

About the Authors

C.E. Dickerson is Professor and Chair of Systems Engineering at Loughborough University in the United Kingdom. His appointment as the Royal Academy of Engineering Chair of Systems Engineering in 2007 began a transition into a sharply focused programme of research after 25 years' experience in aerospace and defence systems in the United States.

His research has aimed to establish a scientific basis for systems engineering supported by viable methods that have utility to industry. The advancements achieved over the years since the appointment are reflected in numerous journal articles published across several fields of the theory and application of architecture and systems. These include complex automotive systems, product line architecture, smart manufacturing, data link interoperability, safety-critical software in commercial nuclear engineering, and formal methods for fault analysis. These advancements have been exploited throughout this book.

His aerospace and defence experience includes the Lockheed Skunkworks and Northrop Advanced Systems in the United States. He was a member of the Research Staff in the Lincoln Laboratory at the Massachusetts Institute of Technology. He has served secondments as the Aegis Systems Engineer for the US Navy Ballistic Missile Defense Program and then as the Director of Architecture for the Chief Engineer in the Office of the Assistant Secretary of the Navy for Research, Development, and Acquisition. Before coming to the UK, he was a Technical Fellow with BAE Systems in the United States.

Professor Dickerson is a Senior Member of the IEEE and Chairs the Mathematical Formalisms Special Interest Group in the Object Management Group. He has also been active with the International Council on Systems Engineering for the past 20 years, to include chairing the Architecture Work Group and later serving as an Assistant Director of Technical Operations. He is also a Fellow of the Higher Education Academy.

Siyuan Ji is a Lecturer of Safety-Critical Systems Engineering in the Department of Computer Science at the University of York in the United Kingdom. His research and teaching are focused on model-based safety assessment and assurance. In his academic duties, he has worked closely with both the aerospace and automotive industries to include BAE Systems and Jaguar Land Rover. He also works closely with the Object Management Group and was the task force lead for the UML Profile for ROSETTA (UPR) that was adopted in 2018. The Relational Oriented Systems Engineering Technology Trade-off and Analysis framework (ROSETTA) is foundational to the methodologies in this book. Dr Ji is also an active member of the IEEE with several publications in the *IEEE Systems Journal*.

Born in China and having been inspired by his grandfather, Fengling Jiang, and the work of physicists like Stephen Hawking, he came to the United Kingdom to study physics at the University of Nottingham where he received his PhD in quantum physics. It was at Nottingham that he also met his lovely wife, Jingjing. His broader interests in science and its application to engineering led him to come to Loughborough University where he joined the research group of Professor Dickerson as a Research Associate. He quickly became actively involved in delivering the post-graduate modules on System Architecture, Design, and Verification upon which the material in this book is based. He is a Fellow of the Higher Education Academy.

List of Abbreviations

ADAS	Advanced driver assistance system
AS	Actuator System
ATC	Air traffic control
ATCC	ATC controller
ATCS	Air traffic control system
ATM	Automated teller machine
ATMS	Air traffic management system
ATR	Air traffic radar
ATSoS	Air traffic system of systems
BSI	British Standards Institute
CAN	Controller area network
CASE	Computer-Aided Software Engineering
CI	Configuration item
CO	Carbon monoxide
CO₂	Carbon dioxide
CPD	Continuing Professional Development
DoS	Disruption of Service
ECA	Electronics Components Association
ECG	Emissions control governor
EIA	Electronics Industries Alliance
IEC	International Electrotechnical Commission
IEEE	International Electronics & Electrical Engineering Association
IET	Institution of Engineering and Technology
EPRSC	Engineering and Physical Sciences Research Council
E–R	Entity–Relationship
HE	Higher Education
INCOSE	International Council on Systems Engineering
IRMS	Intelligent Ramp Metering System
ISO	International Organisation of Standards
IT	Information technology
IUK	Innovate UK
JLR	Jaguar Land Rover
JPL	California Institute of Technology Jet Propulsion Laboratory
MARTE	Modeling and Analysis of Real-time and Embedded Systems
MBSE	Model-based systems engineering
MDA®	Model Driven Architecture®
MIT	Massachusetts Institute of Technology
NCA	Non-compliant aircraft
Nm	Newton-meters
NOx	Nitrogen oxides
OEM	Original equipment manufacturer
OMG™	Object Management Group™

OOSEM	Object-Oriented Systems Engineering Methodology
OPD	Object-Process Diagram
OPM	Object Process Methodology
OSS	Online Shopping System
OZEV	Office for Zero Emissions Vehicles
PIM	Platform-independent model
PIN	Personal identification number
PSi	Programme for Simulation Innovation
PSM	Platform-specific model
RAEng	Royal Academy of Engineering
RFI	Request for information
RFP	Request for proposal
ROSETTA	Relational Oriented Systems Engineering Technology Trade-off and Analysis
RPM	Revolutions per minute
RUP	Rational Unified Process
SAS	System of Actuator Systems
SoS	System of Systems
SysML	Systems Modeling Language
TCC	Traffic Control Centre
TMSoS	Traffic Management System of Systems
UML	Unified Modeling Language
UPR	UML Profile of ROSETTA

Principles of Architecture and Systems Engineering

Conceptual Integrity

Conceptual integrity is the most important consideration in system design.

Role of the Architect

The architect should be responsible for all aspects of conceptual integrity of the system perceivable by the user and understandable by engineers in the development team.

Principle of Definition

Formal definition of concepts is needed for precision; prose definition for comprehensibility.

Structured Analysis

Separation of concerns; the concepts of a problem should be separated from the solution.

Structured Design

Systems should be comprised of loosely coupled highly cohesive modules.

Reflection of Structure

A system solution should reflect the inherent structure of the problem to be solved; in particular, the fundamental form of a system should reflect its intended functionality.

Model Transformation

Model transformations should preserve the semantics and relations of model elements.

Part I

Architecture and Systems
in the 21st Century

1 Introduction

KEY CONCEPTS

Architecting
Systems
Utility
Conceptual integrity

Over a decade ago when the systems engineering community was rethinking its concepts and practices, an expansive advancement of systems engineering tools and languages began picking up momentum. Dating further back to the mid-1990s, the influence of information technology (IT) has become increasingly significant. Object-oriented languages and tools from the IT community were developed and integrated with systems processes and methods. Over the past decade, this integration has been associated with what is broadly referred to as model-based systems engineering (MBSE). Amongst other things, this was fuelled by and supported the development of large-scale information systems in aerospace and defence such as theatre level mission planning systems in which software, hardware, databases, system operators, and increasingly autonomous vehicles are integrated. The role of system architecting became more central and the need for more formal approaches to system specification, analysis, design, and assurance became central to MBSE. This was echoed in the INCOSE Systems Engineering Vision 2020 expressed in 2007 as noted in the INCOSE *Systems Engineering Handbook* (INCOSE 2015).

Many of the tools and modelling languages of MBSE have been standardised by the Object Management Group (OMG). The established object-oriented language for software development, the Unified Model Language (UML), was adapted and has evolved into the Systems Modeling Language (SysML). Concurrently the standards for software and systems engineering continued to evolve. The need for agreed-upon language is a part of these standards. In the case of the term *system architecture*, driven by the need for agreement, new standards for terminology and concepts were initiated. Given the extent and breadth of activity in the systems engineering community over the past decade, it is important to step back and look at architecting and the engineering of systems in the context of the first two decades of the twenty-first century.

This chapter provides 'brief histories' that serve to distil some of the key activities and concepts and speak to some of the key issues. These derive from the INCOSE *Systems Engineering Handbook* (INCOSE 2015), especially the contribution by Estefan (2007); work that includes a critical analysis of MBSE (Dickerson and Mavris 2013), and summative research for using mathematical model theory in architecture definition and design (Dickerson et al. 2021). One purpose of the summative

DOI: 10.1201/9781003213635-1

research was to offer definitions of *architecture* and *system* that are complemented by a mathematical interpretation that can be: (i) used for specification of a technical process for architecture definition and (ii) exploited in engineering practice and relevant standards. The details of the process and mathematically based methods for its implementation are subjects of Chapters 3 and 4 of this book.

What should be clear from the history that follows is that MBSE is far from being in a final stage of maturity. Amongst the areas of work to be finished include a formal basis for MBSE (both logical and scientific) and a stronger synthesis between systems and software engineering.

1.1   A BRIEF HISTORY OF ARCHITECTURE

Architecture is key to the modern practice of engineering. The term would be understood by a general audience as a property of buildings or large-scale structures. In civil engineering, it relates a building's purpose (function), form (how its spaces are organised to achieve the purpose), and construction (what it is built from and how it is built). Although understanding architecture in this way is intuitive and useful, it lacks the precision needed for application to engineering problems. Beyond the design and construction of buildings, architectural ideas are also prevalent in disciplines such as software and systems engineering, management science, and biology.

Despite an extensive common ground, there has been no consensus on the terminology or meaning of architecture across the disciplines. In many ways, a precise practical definition has been elusive. Achieving a precise agreed-upon definition by means of a formal approach could bring unity. The International Organisation of Standards (ISO) established the working group JTC1/SC7/WG42 on System Architecture for this purpose. In the standard for Architecture Description, ISO/IEC/IEEE 42010:2011, a definition was adopted that has been subsumed into later standards (ISO 2011). In 2018, the working group began a routine review of the standard, which is still ongoing at the time of writing of this book. In a shift from a narrow system orientation, recent efforts within have also applied the term architecture to entities not normally considered to be systems.

The early attempts to standardise ideas about the meaning of architecture within systems engineering were in connection with the lifecycle processes standard, ISO/IEC 15288. The first published version (ISO 2002) was influenced by legacy software engineering standards, which conditioned the way in which architecture was conceived. It also combined architecture with design into a single process called Architectural Design. This was separated into two processes in the 2015 update, ISO/IEC/IEEE 15288:2015 (ISO 2015), which is generally accepted as the *de facto* standard for system life cycle development. It has also sought to incorporate the influences and standards from software engineering.

As noted by Wilkinson in Dickerson et al. (2021), the influences of the software discipline date back to the early 1990s, when the IEEE considered architecture to be "the organizational structure of a system or component". This provided a formulation for architecture as a system property, described in terms of static structure (elements and their relations). In 1996, with the aim of using an architectural metaphor as a foundation for systems engineering, Hilliard, Rice and Schwarm (1996) defined architecture as "the highest level conception of a system in its environment". This has several important ideas: architecture is 'high level', includes an 'external focus', and is a 'conception' in the imagination. The ideas described by Hilliard et al. were carried over into the definitions used in IEEE 1471:2000 (IEEE 2000), which defined architecture as "the fundamental organization of a system embodied in its components, their relationships to each other, and to the environment and the principles guiding its design and evolution". The same wording was used in the 2008 issue of ISO/IEC/IEEE 15288. The idea that architecture is subjective is fully embraced in these standards, which introduce architectural descriptions (i.e., models of architecture) as the vehicle for conveying information about the subjective conception of architecture. However, the role of models of architecture for communication has caused a widespread perception that architecture and model are equivalent: Wasson (2005), for example, considers architecture as a graphical model (or representation). By contrast, the popular term 'architecture description' can be understood as referring to a description of a system, just as the term 'blueprint' in civil engineering refers to a blueprint of a building and not of the architecture.

Over the same two-decade period, the Object Management Group™ (OMG™) has used architecture in standards for software development. For example, the Model Driven Architecture® (MDA®) is an approach to software design, development, and implementation. MDA provides guidelines for structuring software specifications that are expressed as models (OMG 2014). In MDA, the software architecture of an application is the basis for deploying the computer code. It provides a comprehensive collection of models that can be used to specify both the software and the computing platform. In the MDA approach, software development focuses first on the functionality and behaviour of a system by means of a platform-independent model (PIM) that separates business and application logic from underlying platform technology. In this way, functionality and behaviour are modelled once and only once. The PIM is implemented by one or more platform-specific models (PSMs) and sets of interface definitions until the specification of the system implementation is complete. A PSM combines the specifications in the PIM with details about how a system uses a particular platform type, but not the details necessary to implement the system.

Despite the fact that MDA has a two decade long standing as an international OMG standard and that the OMG has played a substantial role in MBSE through the development of tools and modelling languages, the concepts of MDA have not played a significant role in the ISO standards related to architecture and systems engineering. This possibly could reflect what appears to be an emphasis on the hardware aspects of systems engineering over those of software in the development of MBSE concepts and practices over the past decade. Another factor might be the lack of a process for specifying the software architecture of modern systems concurrently with the hardware architecture from a comprehensive unified architecture of the system that is attractive to software developers.

1.2   A BRIEF HISTORY OF MODEL-BASED SYSTEMS ENGINEERING

The formalisation of system concepts and processes, and the benefits of so doing have been long recognised in, for example: (i) the foundational work of the biologist Karl L. von Bertalanffy (1967), which was inclusive of many mathematical expressions of systems concepts; (ii) Wymore's codification of *Model-based Systems Engineering* (Wymore 1993), which expressed a programme for systems engineering; and (iii) the mathematical biologist Robert Rosen (1993), who understood the limitations of models and was perhaps the first to recognise the possibility of using Category Theory for systems and scientific problem-solving. Dickerson has also long recognised and investigated formal approaches. The first academic publication with regard to systems concepts (Dickerson 2008) offered an interpretation of the definition of *system* adopted in the standards as a Hamiltonian system in physics to demonstrate that fundamental system concepts could be put on a logical and scientific footing.

1.2.1 EARLY HISTORY OF MBSE

The early history of modelling in systems engineering and key contributions to the formalisation and modelling for software and systems trace back to the second half of the twentieth century. The mathematical concept that a model can be regarded as a nonempty set upon which relations are defined can be traced back as early as Tarski (1954; 1955). A system has also been regarded as a collection of objects with attributes and relations between the objects as well as relations between the attributes and between the relations. Lin (1999) compiled an extensive survey of the mathematical concepts and literature. Lin and Ma (1987) defined a general system to be an ordered pair (M, R) where M is a set and R is a collection of relations on M, i.e., a relational structure. However, regarding a system as a relational structure invites confusion between the concept of a system and that of a first-order model, as defined mathematically by Tarski. Their definition might be suitable for the concept of a system that was later expressed, e.g., in ISO 15288 (2015) as a combination of interacting elements. However, it does not formalise the concept that a system is a "whole" consisting of "interrelated parts", which was expressed by Bertalanffy (1967). This issue is a subject of concern in the book. A resolution and deeper explanation of the issue is rooted in the critical analysis of the definitions of key terminology in Chapter 3. A final logical model of the concepts of architecture and system is offered in the summary at the end of Chapter 4.

There is also a significant distinction to be made between the concept of interrelated elements and that of interacting elements. Interaction is very much a concept from physics whereas interrelationship is a concept more common to domains of knowledge such as biology and economics. Furthermore, the mathematical concept of relation as used in first-order model theory *does not per se* address relations between relations, which is a fundamental conceptualisation of interrelationship. Bertalanffy persistently sought a formalisation of the concept of interrelationship but never found one to his satisfaction.

With the emergence of digital computers during this period, Yourdon (1989) introduced a model-based approach for software development from a behavioural viewpoint. He also introduced a graphical modelling language called Data Flow Diagrams to support his approach. His concept of Structured Design was based on a principle that systems should be comprised of modules, each of which is highly cohesive but collectively are loosely coupled. The strongest form of cohesion is based on functionality. The idea was that cohesion within modules minimises interactions between elements not in the same module; thus, minimising the number of connections and the amount of coupling between modules.

Yourdon also introduced the concept of Structured Analysis, which is based on a principle that the specification of the problem should be separated from that of the solution. He proposed that this separation be accomplished by using two types of models: an Essential Model, which is an implementation free representation of the system, and an Implementation Model, which is a specific representation of the software and hardware needed to realise the system. These two types of models are linked by interpretation of a behavioural model into one for implementation. Structured Analysis requires that the concerns of the two types of models be separated. These concepts for using models in software development became the basis for the OMG MDA approach. The concepts and models of Structured Analysis and Design will be explored further in Chapter 4 then carried forward throughout this book.

Wayne Wymore (1993) was one of the early pioneers in the domain of model-based systems engineering. He had a behavioural viewpoint on systems in which the model of behaviour is comprised of the name of the system, its states, the inputs and outputs of those states, and state transitions. A readout function was also provided for outputs of the transitions. Wymore advocated that system design is the development of a model on the basis of which a real system can be built, developed, or deployed that specifies all the requirements using a mathematically based system design language.

Wymore also had a concept of homomorphism between system models as a mapping of the states of the systems, to include their inputs and outputs, in such a way that the two homomorphic models exhibit the same behaviour under the mapping. Homomorphism between functional and implementation system models was intended to assure system intended behaviour. This is similar to Yourdon's concept of Structured Design that the solution should reflect the inherent structure of the problem.

Wymore's concept of homomorphism is generalised in Lin (1999) by the concept of a general system using an algebraic definition. Specifically, given two general system models $S_1 = (M_1, R_1)$ and $S_2 = (M_2, R_2)$, a mapping $h: M_1 \rightarrow M_2$ is a (relational) homomorphism if for each relation $r \in R_1, h(r) \in R_2$. Note that this definition preserves relations at a structural level of description. It should also be noted that (i) in abstract algebra the mapping h is a function and is not permitted to make multi-valued assignments into S_2, and (ii) as noted previously, this risks confusion between the concept of a system and that of a model. The mathematical details of this concept will be clarified in Chapter 2 (Logical and Scientific Approach), then used extensively through the book.

Towards the end of the twentieth century, Klir (1991) complemented the concept of a general system with a general systems methodology. Simply stated, he regarded problem solving in general to rest upon a principle of alternatively using abstraction and interpretation to solve a problem. He considered that this could be used both for system inquiry (e.g., the modelling of an aspect of reality) and for system design (i.e., the modelling of purposeful human-made objects). The Klir concept of system inquiry can also be regarded as an approach to architecture and system description as considered in the twenty-first century standards. This will be exploited extensively in the book beginning with Chapter 4 when a relation framework is introduced that implements the concepts of homomorphism expressed by Wymore and Lin.

Looking back on the second half of the past century, key historical contributions were made to establish model-based approaches for system description, analysis, and design. The concepts were strongly influenced by software engineering but were multidisciplinary and sought to establish systems as a domain of knowledge founded on mathematics and science. Generally, a system was regarded as a collection of objects (entities) with attributes and relations between the objects, relations between their attributes, as well as interrelations. A mathematical formalisation of system modelling was consistently sought.

In principle, the formalisation could have been accomplished using the first-order model theory of Tarski and the homomorphism of relational structures prescribed by Klir and Lin. Each of these model-based approaches in the early history of MBSE incorporated a mathematical form of model transformation that preserves structure and behaviour. However, the systems community never achieved closure at this level of formalisation. In the current practice of MBSE over the past decade, a less formal and more intuitive approach has been followed in software and systems engineering using graphical object-oriented modelling languages such as UML and SysML. The idea of transformation between models has been part of this approach but has never been formalised and applied with a precision sufficient for engineering applications.

1.2.2   Recent History of MBSE

Developed in the 1990s by the Chesapeake Chapter of INCOSE with significant aerospace involvement, the Object-Oriented Systems Engineering Methodology (OOSEM) is a top-down hybrid approach that leverages object-oriented software and system techniques. It is model-based and can be implemented in SysML. The core tenet is integrated product development using a recursive Systems Engineering Vee approach. Key system development activities in the methodology include: Analysis of Stakeholder Needs, Definition of System Requirements, Definition of the Logical Architecture, Synthesis of Candidate Allocated Architectures, and Optimisation and Evaluation of Alternatives. OOSEM is supported by a number of tools that have been commercially developed using OMG standards and SysML.

This and other methodologies were reviewed by Estefan (2007) in an INCOSE report. One was based on concurrent systems engineering activities that reflect the Systems Engineering Vee: the Vitech MBSE Methodology. The key activities are: Source Requirements Analysis, Functional Behaviour Analysis, Architecture Synthesis, and Design Validation and Verification. The methodology uses a layered approach to system design (the 'Onion Model') and a common System Design Repository. It specifies a System Definition Language based on entities, relationships, and attributes. As noted by Estefan, the methodology is commercially supported by the Vitech CORE product suite.

Another methodology is the State Analysis that was developed by the California Institute of Technology Jet Propulsion Laboratory (JPL) with deep space missions in mind. The methodology uses Goal-directed Operations Engineering and leverages model- and state-based control architecture. States describe the 'condition' of an evolving system, such as a spacecraft over possibly long periods of a mission. This is an iterative process for state discovery and modelling. Models are used to describe system evolution. Core tenants include state-based behavioural modelling and state-based software design. The methodology seeks to reduce gaps in the software implementation of systems engineering requirements. The JPL State Analysis can augment the Vitech CORE functional analysis schema to synthesise functional and state analysis.

A formal modelling methodology was developed by Dori (2016): The Object Process Methodology (OPM), which is based on the premise that everything is either an object or a process. Objects exist or have the potential for existence. Processes are patterns for the transformation of objects. States are situations that objects can be in. OPM combines formal Object-Process Diagrams (OPDs) with an Object-Process Language (OPL). OPD constructs have semantically equivalent OPL sentences. The reader should note the similarity with Tarski model theory, i.e., language is interpreted into structures (diagrams in this case).

OPL is oriented towards humans as well as machines. There are also system structural links (relations) and procedural links (behaviour). Structural links are similar to UML class relationships (e.g., generalisation). Three mechanisms are used for modelling: (i) unfolding/folding that refines/abstracts structural hierarchy, (ii) zooming out/zooming that exposes/hides details of an object or process, and (iii) expressing/suppressing: exposes/hides details of a state.

Estefan also reported on two IBM methodologies: The Rational Telelogic Harmony method and the Rational Unified Process (RUP) for systems engineering, which extends development methods for software to methods for system development. Modelling is supported at the levels of context, analysis, design, and implementation by the IBM Rhapsody Suite.

The Estefan survey is representative of the progress in MBSE over the past decade but is not exhaustive. What is evident though from the recent history is that both a functional orientation and an object orientation have been carried forward that are embodied in processes, methods, and languages that are supported by commercially available tools. Formal methods have also progressed. However, the vision of the twentieth-century pioneers remains to be completely fulfilled by a comprehensive approach with a logical and scientific basis that expresses the core concepts of MBSE.

1.3 INVENTION AND UTILITY

Thomas Edison was a noteworthy inventor in the US during the early twentieth century. Remembered for inventing the electric light bulb, he became a wealthy and recognised technology leader of his time. His pursuit of invention could be considered the pursuit of valuable intellectual capital. What was the secret to his success and how did he discover it?

At the age of 22, Edison received his first patent: an electronic vote recorder to be used by legislative bodies. It was a simple concept but very advanced for its time. Votes could be recorded by the flip of a switch, but legislators did not trust this system and preferred to cast their votes by voice. Edison could not sell the invention. The inventor's response was:

> "Anything that won't sell, I don't want to invent. Its sale is a proof of utility, and utility is success."

In the years to come, he applied this principle to his engineering practice of invention. He became very wealthy and some would say that he gave up his passion for technology in the pursuit of money. But another view of Thomas Edison might consider that he valued the utility of his inventions and focused on the utility of the technologies emerging in his day rather than on just the technologies. His viewpoint coupled with his passion for technology could not help but make him a very wealthy businessman. Even so, Edison himself said that invention was 99% perspiration and 1% inspiration, and it has been said of him that often he was found late at night sleeping on one of the laboratory benches, in between running experiments.

1.3.1 THE UTILITY OF SYSTEMS ENGINEERING

Engineering was well-established across many scientific disciplines by the beginning of the twentieth century: mechanical, electrical, and chemical, just to name a few. Engineers have since established themselves as the twenty-first century applied scientists and like Edison deliver technical products and services of value. It is a legitimate question to ask how well systems engineers have established themselves as they now enter the third decade of the twenty-first century. The INCOSE *Systems Engineering Handbook* speaks to the value and offers studies and analyses (INCOSE 2015), though many are large-scale and defence-oriented.

There is a general recognition that systems engineering is important for discovering problems early in the development of products, thereby dramatically reducing the cost of scrap and rework in the later stages, or in the worst case, program cancellation. In fact, the leading cause of cancellation for large-scale software-intensive systems development is the failure to capture or manage requirements. Surveys conducted by INCOSE give further insight relevant to MBSE. The results presented by Cloutier and Bone (2015) indicated that most practitioners were primarily performing the technical processes of Requirements Definition and Architecture Definition (as specified in ISO/IEC/IEEE 15288: 2015). This should not come as a surprise. Two decades ago, Hatley's *Process for Systems Architecture and Requirements Engineering* (Hatley, Hruschka and Pirbhai 2000) was focused entirely on these two technical processes as based on a successful commercial track record in the 1990s. The actual business value of MBSE is a longer term proposition in that the cost savings in a shift from document- to model-based management of information nominally is realised only as the product or service enters production and deployment.

1.3.2 THE UPTAKE OF MBSE

There are various sources regarding how well industry as a whole has adopted the processes, tools, and modelling languages of MBSE. In the United States, a recent survey was conducted by the Massachusetts Institute of Technology (Cameron and Adsit 2020). This survey of documentation methods was conducted across several thousand post-graduate students enrolled in online studies at MIT. The survey only reflects the usage of methods at the student's company. It should also be noted that the students were mostly mechanical engineers whose companies did not have an MBSE process in place. It is therefore possible that some or many of the students were enrolled because of corporate interest to explore MBSE possibilities.

Most of the documentation methods reported in the survey were concerned with requirements (slightly fewer than 2,000 responses) and change requests (about 800 responses). Word file usage was the most frequent method in both cases, accounting for about one-third of the responses for requirements documentation, which was comparable to the number of responses that reported using software tools for requirements documentation. Word files accounted for almost two-thirds of the responses reported for change requests. For interface control (about 600 responses), the usage of software tools was less than one-third. Word files accounted for over two-thirds. There were just fewer than 500 responses collectively for UML, SysML, Design Structure Matrix, and OPM; with UML being the most frequent at about one-third of the responses and SysML slightly less.

Reasons for usage were not analysed in the survey. Nonetheless, the scale of preference towards text-based documents for requirements, change requests, and interface control over UML, SysML, and OPM is dramatic. It is also worth noting that UML has not been displaced by SysML in the companies in the survey. This might reflect that software engineers have been using UML for software development for over two decades whereas SysML is more recent; or, on the other hand, that they generally prefer to use UML. The survey is silent on such details.

This is perhaps an appropriate point to reflect on the context and progress in systems engineering that has been made since the first edition of the book, which was written over a decade ago when the community was rethinking systems engineering. There is no definitive research on the benefits and utility of the progress that has been made during the expansive advancement of systems engineering tools and languages over the past decade. This is not to say that progress has not been made. However, research like that of MIT would indicate that an overwhelming impact over the past decade of advancement has not been achieved.

The words of Edison that "utility is success" are perhaps the end goal that the systems engineering community should keep in mind at the beginning of the next decade of advancement that lies ahead.

1.4 THE ROLE OF THE ARCHITECT

Finally, it is worth considering how the system architect fits into this picture. In practice, a system architect has a more user and customer-focused role, whereas a systems engineer might be more concerned with the technical team developing the product or service. In a small-scale project, the functions of the architect and systems engineer may indeed reside in the same person. This is certainly the case in civil engineering. A pioneer of commercial digital computer architecture, F. P. Brooks (1995) succinctly expressed a viewpoint on the role of the architect:

> "Conceptual integrity is the most important consideration in system design. The architect should be responsible for the conceptual integrity of all aspects of the product perceivable by the user."

These two statements have been taken as the first and second principles of architecture and systems engineering in this book. This viewpoint is supported by the professional experience of the principal author, tracing as far back to his role as the Director of Architecture for the Chief Engineer of the U.S. Navy (2000–2003) in the Office of the Assistant Secretary of the Navy for Research, Development, and Acquisition. In this role (Dickerson et al. 2003), the architect was positioned between the acquisition authority (the Assistant Secretary) and the government leaders of system development (the System Commands). This is not unlike the position that many system architects in aerospace and commercial practice find themselves in today. This viewpoint was taken in the first edition of this book and it remains unchanged a decade later.

2 Logical and Scientific Approach

KEY CONCEPTS

Conceptual integrity
Knowledge and language
Models in science and engineering
Logic and first-order models

Mathematics is the language of science and engineering. However, neither necessarily uses the formal language upon which mathematics is founded in a formal way. Instead, like many other domains of discourse, their use of language is more descriptive of concepts that have a physical foundation. The validity of these domains is based on repeatable experiments or the successful realisation of concepts in terms of physical objects. Modern mathematics and science enjoy the benefit of thousands of years of human thought, experience, and practice. The professional practice of engineering, spurred on by the industrial revolution, has enjoyed centuries of such benefits. Systems engineering, on the other hand, has enjoyed less than a century of development as a professional practice. As a distinct engineering discipline, from any historical perspective, it must be considered to be still young and maturing.

Systems engineering by its very nature is multidisciplinary. The use of language in this developing domain of discourse is then particularly challenging because it needs a common language to express both domain concepts and system concepts. The technical use of the language of mathematics prevalent in science and engineering has not been developed for systems to date in a way that has been applied across multiple domains (engineering disciplines) to enable a coherent and comprehensive approach to the engineering of both hardware and software for modern products and services, i.e., systems. This chapter begins a discourse on how the language of logic binds mathematics, science, engineering, systems, and software together.

Comparing mathematics and science with the historical and academic disciplines of engineering, it is clear that:

- Mathematics uses
  - Formal logic (the predicate calculus of logic) for description
  - Logical deduction and mathematical induction for reasoning
- Science uses
  - Mathematics for description and quantification
  - Models and experimental methods for reasoning

DOI: 10.1201/9781003213635-2

- Engineering disciplines use
  - Science and mathematics for description and reasoning
  - Analysis, tests, and prototypes for design and decision
- Software engineering uses languages and tools
  - Computer languages such as the Language C for procedural programming
  - The Unified Modeling Language (UML) for object-oriented programming
  - Computer-Aided Software Engineering (CASE) tools

The separation of (hardware) engineering from software engineering in the above comparison is necessary because these two disciplines are distinct domains of discourse. This distinction raises many issues that are a subject of research. An investigation of the issues, which is accessible to general audience, can be found in the work by Pyster et al. (2015). How architecture and systems can better unify these two domains is a subject of concern throughout this edition of the book.

2.1 EXPERIMENTAL AND LOGICAL BASIS OF SCIENCE

If engineering in general and systems engineering in particular are to be founded upon science, then there needs to be an agreement on what is meant by *science*. The following perspective is offered as a fundamental understanding of the term.

- The Latin word *Scientia* means knowledge.
- The English word *science* refers to any systematic body of knowledge but more commonly refers to one that is based on the scientific method.
- The scientific method consists of
  - Characterisation of observables (by definition and measurement)
  - Formulation of hypotheses (e.g., interpretations of models) which are testable and refutable by comparing:
    - Predictions (based on the hypotheses) and
    - Outcomes of experiments (which must be repeatable)

It will be important to understand the difference between principles and laws in both science and systems. In physics, principles are deeper insights from which laws can be derived. Consider the following:

- Examples of principles
  - The conservation of energy
  - The equivalence of mass and energy
- Examples of laws
  - Newton's Laws of Dynamics
  - The Law of Gravity

The laws of Newtonian dynamics, for example, can be derived from the principle of the conservation of energy. Science generally seeks to develop mathematical models of its laws. Consider Newton's three laws for describing the dynamics of bodies of matter.

- Law of Inertia: *An object will stay at rest or continue at a constant velocity unless acted upon by an external force.*

- Law of Acceleration: *The net force on an object is equal to the mass of the object multiplied by its acceleration.*
- Law of Reciprocal Actions: *Every action has an equal and opposite reaction.*

These laws are given in natural language but can be modelled using algebra and calculus. The power of Newton's calculus allowed many physical systems to be modelled and their behaviour predicted. The simple Law of Acceleration, when properly modelled with Newton's calculus, leads to Hamiltonian mechanics where systems of large numbers of particles or other physical objects are described by partial differential equations. The Hamiltonian classical theory of mechanics developed in the nineteenth century is surprisingly universal. It can be used, for example, to formulate models in quantum mechanics, a domain of physics developed in the twentieth century that dispelled many classical concepts.

If this is sounding too easy, you're right! What makes it hard is that a *problem* is easier to model than finding a *solution*! For example, the well-known equations for a system of three bodies of mass remain unsolved even today. The orbits of the planets in our solar system would be unpredictable if it were not for the dominant mass of the sun!

So, the maturing discipline of systems engineering has something to learn from the historical disciplines of science. The power of formal languages in modelling is just one. But a new level of precision for the description of systems engineering problems will not be a *panacea*. Hard problems and unsolvable problems will remain. Reliable prediction of behaviour is not always attainable.

2.2   SCIENTIFIC BASIS OF ENGINEERING

If systems engineering is to be properly understood, then it must be understood in the general context of engineering. The following definition of engineering will be used for this book:

> *Engineering is a practice of concept realisation in which relations between structure and functionality are modelled using the laws of science for the purpose of solving a problem or exploiting an opportunity.*

The fundamental nature of this definition will become clear throughout its use in the book. Although there are no doubt many other useful definitions of this common term, this definition is well suited to the architecture and systems engineering approach that will be taken.

2.2.1   STRUCTURAL ENGINEERING EXAMPLE OF USING SCIENCE

A simple example from structural engineering can be used to illustrate the relation between science and engineering:

- Problem: how to best arrange bricks to build a bridge
- Purpose: the bridge enables transport across a gap in the terrain

- The underlying physics
  - Law of Reciprocal Actions
  - Principle of Dispersion of Force in a Structure
- Implications for the engineering of the bridge
  - The bridge must transfer sufficiently large forces from the plane of transport to the base of the bridge on the ground.
  - If the bricks are arranged in an arch, then the forces can be evenly dispersed through the structure and transferred to the base.
  - The layers of bricks should be staggered to avoid vertical seams in the mortar. (Such seams would provide paths of stress in the structure through the weaker material that comprises the mortar.)
  - Among the benefits of the design is that the number of bricks required is greatly reduced and the resulting structure can be visually attractive.

It should also be noted that the size of the bridge might be limited by the size of the bricks.

Civilisations have been building bridges such as this for thousands of years without the benefit of modern science. Only a rudimentary understanding of what makes a structure stable and durable was needed. The arch is a *type of structure* that has long been recognised as a preferred arrangement of construction materials (stones, bricks, etc.). Historically the designer might have been an *architect* who would work out various details as to how to build the bridge using this type of structure in a way that best achieved the purpose, i.e., to enable transport across a gap in the terrain and have other properties such as durability and attractiveness.

This short intuitive description of engineering should not lull the reader into a false sense that good engineering comes easily. Notwithstanding the commercial and other contextual problems that can challenge the engineer, reliably understanding and predicting the relationship between functionality and physical structure is a serious technical problem. The laws of science can work both for engineers as well as against them. Murphy's Law is always at work.

The three-body problem in physics is not an isolated example of an inability to predict behaviour. This inability can be complemented by a lack of understanding of the environment of a system. The catastrophic collapse of the Tacoma Narrows Bridge in 1940 stands as one of the great examples. Harmonic oscillations that are well understood in science and engineering were the cause of its collapse due to a lack of understanding of the wind flows in the locale of the bridge. More recently, the Millennium Bridge in London suffered a similar design flaw, which fortunately was corrected before a catastrophic event occurred (but only after the bridge went into service). The unforeseen harmonics, in this case, were caused by large numbers of pedestrians on the bridge who had self-synchronised their cadence. Thus, the use of science in building modern bridges becomes a necessity rather than simply an interesting insight into the behaviour of bridges.

2.2.2 EXAMPLE OF USING SCIENCE IN SYSTEMS ENGINEERING

A more detailed technical example is provided by twenty-first century automotive engineering. The problem in this example will be how to control engine emissions while simultaneously providing cruise control functionality. Advanced driver assistance systems (ADAS) are a class of information-intensive systems that can be developed using science and engineering models to produce solutions that improve driver comfort and safety. A legacy ADAS functionality that has evolved for decades is realised in cruise control systems. In its most basic form, the system maintains a constant speed for the vehicle but gives the driver an option to accelerate the vehicle any time by simply pressing the accelerator pedal. This example will illustrate the relation between science and engineering as follows:

- Problem: how best to limit internal combustion engine emissions during cruise control
- Purpose: maintain driver comfort and control whilst complying with emissions standards
- The underlying science:
  - Law of Inertia
  - Law of Acceleration
  - The chemistry of internal combustion engines
- Implications for the engineering of the ADAS and vehicle engine:
  - A mapping is needed of engine states versus emissions
  - Operating ranges within the permissible states must be robust enough to
    - permit the ADAS to maintain a constant vehicle speed and
    - meet driver demands for vehicle acceleration

One particular challenge in this class of problems is that despite the advances of a century of study of the chemistry of internal combustion engines, it is currently not possible to model or simulate engine behaviours at a level of fidelity that can be used to reliably predict compliance with emissions standards. Measured data from an engine is the basis for creating a reference model that is referred to as an engine map.

A coordinated control architecture design (Lin et al. 2018) can be applied to this class of emissions control problems. The concept is that the existing architecture of the ADAS control of the engine will not be altered but instead a separate emissions control function that acts as a governor will be interfaced with the existing cruise control function and the engine control unit. One advantage of this approach is that the modular system architecture of the integrated solution will not require redesign or extensive retesting of the existing systems.

For the sake of simplicity of illustration, only two design objective constraints will be specified for emissions and only two state variables for the engine. The actual problem has dozens of variables and constraints. This problem was a subject of research in a 5-year programme sponsored by the UK Research Councils as summarised by Dickerson and Ji (2018), the results of which led to a patent on emissions control (Dickerson, Ji and Battersby 2018) amongst other advancements.

In this simplified problem, the two objectives are $z_1 = CO_2$ emission measured in kg/hour and $z_2 = CO$ emission measured in kg/hour. The constraints are $z_1 \leq 30$ and $z_2 \leq 0.1$. The state variables will be engine speed, x_1 in revolutions per minute (RPM) and engine torque, x_2 in Newton-meters (Nm). An engine map has been measured for the state space $1600 \leq x_1 \leq 2000$ and $100 \leq x_2 \leq 240$. As a mathematical mapping, the map is defined by two response surfaces $f(x_1, x_2) = z_1$ and $g(x_1, x_2) = z_2$.

However, as already noted, the actual engine map is a table of discrete measurements. The selection of measurement points for creating the map is guided by an understanding of the chemistry of internal combustion engines within the intended operating ranges (states) of the engine.

The constrained operating space of the engine displayed by the shaded region of the graph in Figure 2.1(c) depicts the transformation of the constraints on the objectives into engine torque and engine speed through the inverse mappings of the two response surfaces that are displayed in Figure 2.1(a) and (b), respectively. The existing ADAS cruise-control functionality provides torque demands to the engine to maintain a constant vehicle speed (which has been set by the driver). In a fixed gear, this corresponds to a constant engine speed, e.g., 1600 RPM. In this case, the graph in Figure 2.1(c) indicates that the existing ADAS functionality could make torque demands on the engine over the full operating range (100–240 Nm) without violating the emissions constraints. The control of the engine speed by the ADAS would be sufficient for maintaining an appropriate vehicle speed (corresponding to engine speeds near 1600 RPM) and emissions.

At a higher engine speed, e.g., 1700 RPM, the torque demand would need to be limited to about 230 Nm. Thus, the new functionality is indeed acting as a governor on the torque demand permitted by the ADAS. At engine speeds of more than 1700 RPM, the torque demand will need to be not just limited by an upper bound, but it will also need to stay above a lower limit. For example, at 1800 RPM torque demand would need to be limited to be between 130 and 220 Nm. If a torque less than 130 Nm is needed to maintain the vehicle speed that corresponds to 1800 RPM, then the ADAS or the driver will need to shift gears to lower the engine speed. In general, on a smooth level road surface in light traffic conditions, due to inertia, the ADAS can maintain a constant vehicle speed by making small demands for torque to overcome friction and wind resistance. If the vehicle is travelling over hills, a larger amount of torque will need to be demanded.

The governor function for permitting the driver to accelerate from a state of cruising is similar but more complicated. At the start of acceleration action from a constant speed cruise, the engine is initially at a given state (RPM, torque) within the constrained operating space of the engine displayed by the shaded region of the

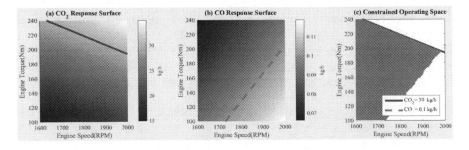

FIGURE 2.1 Engine emissions response surface model and constraints: (a) CO_2 response surface, (b) CO response surface, and (c) constrained operating space.

graph in Figure 2.1(c). When the demand for increased torque is sent to the engine, the result will be increased RPM, i.e., acceleration. Keeping the engine state within the permissible range of states during the acceleration can again be accomplished by limiting the torque demand but the engine speed must also be limited to a permissible range. Thus, the governor must be able to limit the engine speed as well as the torque demand. This could be accomplished by reducing the torque demand permitted to stop the acceleration. For example, if the driver began accelerating from the engine state of 1600 RPM engine speed at 170 Nm, the torque demand would need to be limited to about 205 Nm and when the acceleration brought the engine close to a speed of 1900 RPM, the torque demand would need to be reduced to a level that stops the acceleration. This point is determined by the CO_2 constraint in the graph in Figure 2.1(c). As with the governance of the constant speed cruise, it may also be necessary to shift gears to keep the engine within the permissible range of states.

The analysis of the constrained operating space to specify operating ranges within permissible engine states provides a straightforward model-based method to integrate an emissions control governor with the ADAS. The implementation of this method will require the specification of real-time software. The governor example will be completed in Chapter 4 using Structured Design.

2.3 LANGUAGE, LOGIC, AND MODELS

If the laws and models of science are to be used as a basis of precision and repeatability in engineering, then criteria for what constitutes a good model need to be considered. For architecture and systems engineering, models for both science and conceptual solutions must be considered.

- Stephen Hawking (1988) attributed a good model in physics to be
  - Simple
  - Mathematically correct
  - Experimentally verifiable
- By analogy, a good logical model of a concept should be
  - Simple
  - Logically well-formed and consistent
  - Verifiable through logical interpretation

Concepts can be expressed in various ways, but in practice, natural language sentences remain the most common way to express concepts. The modelling of concepts in this book is motivated by the mathematical theory of first-order models using the predicate calculus. See, e.g., *Models and Ultraproducts* (Bell and Slomson 1969). The terms *model* and *sentence* have precise meanings in the predicate calculus. The syntax of sentences in mathematical logic is specified by the propositional calculus.

- The formal languages of mathematical logic are the
  - Propositional calculus
  - Predicate calculus

- The term *calculus* derives from the Latin for calculation.
  - In logic, *Calculus* refers to the calculation of truth.
  - By contrast, in the *Calculus* of Newton and Leibniz, the term refers to the calculation of limits.
- *Propositions* are declarative statements and are represented by propositional variables, i.e., the propositions are represented as abstract variables such as, p, q, ...
- *Predicates* are statements of relations and are represented by predicate letters such as, P, Q, ... The variables related by a predicate letter are represented as abstract variables such as, u, v, ... Thus, $P(u, v)$ represents an abstract relation between two variables.
- *Quantification and Interpretation*: given a domain of knowledge, these symbols are interpreted into the domain, i.e., they take on meaning and values in the domain. When each variable is interpreted into either a single value or a specified range of values, the predicate is said to be quantified.

Propositional logic is also referred to as the Sentence Calculus because it defines the syntax of logical sentences. Just as algebra makes general statements about numbers (e.g., $a^2 + b^2 = c^2$) that can have various interpretations which are true or false (if proper syntax is followed), so too statements in logic are sentences formed of abstract variables that represent decidable assertions that are generally formed from relations. The types of relations of primary interest in systems engineering will be constraints and interactions between elements.

The shaded region in Figure 2.1(c) shows a constraint relation between the two state variables in the engine operating space: engine speed, x_1, and engine torque, x_2. Each variable has been quantified by its range of values, $1600 \leq x_1 \leq 2000$ and $100 \leq x_2 \leq 240$. Thus, the abstract predicate $P(u, v)$ can be interpreted as the joint constraint relation in two variables in the figure, and it has been quantified. This is the level of formality that is behind every well-formed mathematical model. In the practice of science and engineering, this formality is suppressed, but if the model does not conform to this formal basis, then it will be logically incorrect and exhibit errors.

This is how first-order model theory is conceived in mathematics. A *sentence* is a well-formed formula of predicates that is fully quantified, i.e., the formula adheres to the syntax of logic and there are no free variables in the formula. Free variables (i.e., variables that have not been given proper scope through quantification) admit unbounded interpretation, which is problematic. In natural language, the scope of terms can be provided by certain types of adjectives, e.g., the speed of an aircraft can be quantified by the adjective subsonic.

For a sentence in the predicate calculus, a *model* (of the sentence) is a relational structure (i.e., a set with a collection of relations on the set) into which the sentence can be interpreted and its validity can be reasoned about. The shaded regions determined by the constraints in Figure 2.1(a) and (b) are an example of a collection of two relations that form this type of structure, i.e., there are two relations in the structure. One is the constraint on engine states from CO_2

regulations and the other is a constraint on the states from CO regulations. They are joined by set intersection (which corresponds to a logical '*and* statement') that forms the constrained space.

The relational structure is said to be a *model of a sentence* if the sentence is valid when interpreted into the structure. This means that the 'image' of the interpretation is 'true'. (The model then represents the sentence.) In natural language, a sentence for Figure 2.1(c) could be as simple as, 'There is an operating space of the engine that is constrained by emissions regulations'. This is intuitive but very ambiguous. In the set-theoretic example given by the graphs in Figure 2.1, the constrained operating space of the engine is a *non-empty set*. If the relations in the figure had an empty intersection, then the statement of constraints would be invalid; it cannot be realised.

This type of model is referred as 'first order' because the predicate calculus is a first-order logic. A general but technical review of the model theory for the non-specialist is given by Wilfrid Hodges in the Stanford Encyclopedia of Philosophy (Hodges 2020). The power of this understanding of models is that it crosses the boundaries of science, natural language, and software engineering. Chapter 3 on concepts, standards, and terminology will conclude with a demonstration of how first-order model theory applies to natural language and the definition of terms. The structured methods of Chapter 4 will show how the theory applies to system specification using modelling languages such as UML.

To the engineer and the scientist, this level of logical detail might seem at best to be an interesting insight into the modelling of systems. Indeed, just as certain details of mathematics are suppressed in science and engineering, many of these logical details should be managed by proper use of natural language and also in the background logic of a modern systems modelling tool using a language such as UML or SysML. However, first-order models are much like the use of science in building modern bridges where the scientific understanding of the behaviour of bridges becomes a necessity rather than simply an interesting insight. Just as bridges can fall down, concepts can fall apart when subjected to the complexity of integrating hardware and software into a holistic system. If conceptual integrity is not maintained, the result is failure to realise the concepts and requirements that the system was intended to fulfil. Indeed, the system might even fail to be built.

3 Concepts, Standards, and Terminology

KEY CONCEPTS

Purpose of engineering standards
Interpretation of terminology
Concepts and model theory
Essential technical processes

The terminology of systems engineering has different meanings depending on the user. Standards organisations seek to normalise terminology across a broad community of users. However, the current standards relevant to systems engineering *do not* provide the precision required by a rigorous model-based approach to architecture and systems engineering. Precise but practical definitions of systems terminology that have collective consistency and conceptual integrity and also have engineering utility are needed for the specification of technical processes and methods for the system and architecture definition of engineering solutions to problems of commercial interest. This chapter introduces precisely defined terminology that can be used to complement the current terminology in accepted standards and to specify processes and methods based on the interpretation of terms.

Beginning with this foundation chapter, a lexicon of terms and definitions is developed that supports the specification of essential technical processes and methods. Their use and utility are demonstrated in the tutorial case study covered by Chapters 5, 6, and 7. A more comprehensive lexicon is provided in Annex A-3. The concepts and terminology used throughout this book are derived primarily from ISO/IEC/IEEE 15288:2015 (ISO 15288) to the extent possible. This is the most recent and most widely adopted international standard on software and systems engineering life cycle processes. This and other standards will be complemented by research findings to establish the authority and correctness of the concepts and terminology in this book.

Based on the first-order model theory of Tarski presented in Chapter 2, the pairing of models with concepts expressed in terms of a language is then both intuitive and mathematically based. This approach demands both simplicity and rigor. The essential terminology and technical processes presented in this book underlie those specified in the relevant international standards for systems engineering. These are based on recent summative research by the authors and their collaborators (Dickerson et al. 2021). It should be noted that the international standards also consider the commercial and life-cycle aspects of engineering such as definition and management of the system configuration, and translation of system definition into work breakdown structures, amongst others. The essential definitions and processes presented in this book are intended to complement the standards; not to be a replacement.

This chapter concludes with a demonstration that concepts expressed by natural language definitions of terminology can be modelled with the same level of formality as the science-based advanced driver-assistance systems example in Chapter 2. The foundation that has been laid for a scientific basis of engineering using first-order model theory will be further demonstrated by modelling a widely accepted definition of the term system. This will be an important step towards establishing that the approach taken in this book is a qualified approach to model-based systems engineering. Chapter 4 will carry this further by showing how the concepts of architecture and systems can be expressed in a logical model that enables engineering through the methods of structured analysis and design. Using the common thread of an underlying logic that realises the science and engineering concepts of the emissions control governor investigated in Chapter 2, a model-based specification of a system solution will then be described to conclude Part I.

3.1 SYSTEMS ENGINEERING STANDARDS AND ORGANISATIONS

When dealing with the standards promulgated through various national and international organisations, it is important to understand that the standardisation of concepts and terminology is an agreement between users, practitioners, and other stakeholders of the standard. As such these standards tend to reflect a consensus of current best practice across differing business and technical domains. Furthermore, an international standard must also account for interpretation into multiple languages. It can easily be the case that a single key word in English, for example, has no corresponding word in one of more of the languages of Europe, much less the many languages and dialects of countries such as India and China.

With so many influences on the choice and use of terminology, it should *not* be expected that the current concepts and terminology of systems engineering have been formalised to a level that bears the test of logical and scientific precision. In the domain of science and mathematics, the use of precise symbolic languages has greatly alleviated this problem, but it has taken centuries for that normalisation to occur. Systems engineering in comparison with science and mathematics is still a young subject that has a significant intellectual distance to go before its concepts and terminology can be normalised to the level of precision and ubiquity that is currently enjoyed by science and mathematics.

In the current state of standards and practice, the key international organisations with published and broadly adopted concepts relevant to architecture and systems engineering are the

- International Organization for Standardisation (ISO)
- British Standards Institute (BSI)
- International Electronics & Electrical Engineering Association (IEEE)
- Institution of Engineering and Technology (IET)
- International Electrotechnical Commission (IEC)
- Electronics Industries Alliance (EIA)
- Electronics Components Association (ECA)
- Object Management Group (OMG)
- International Council on Systems Engineering (INCOSE)

INCOSE is not a standards organisation but rather is an industry focused professional society founded to develop and disseminate the interdisciplinary principles and practices for systems engineering. It does, however, work closely with standards organisations. For example, the INCOSE *Systems Engineering Handbook*, 4th Edition (INCOSE 2015) was published in parallel with the publication of ISO/IEC/IEEE 15288:2015 (ISO 2015), which is the current *de facto* standard on systems and software engineering. The handbook seeks to elaborate the standard for the purpose of industrial practice.

ISO is the largest developer and publisher of a broad range of international standards. The BSI is a national standards body that was awarded a Royal Charter in 1929 and reaches across the international standards community. The IEEE is a non-profit organisation and is a leading professional association for the advancement of technology, publishing not only standards but also a wide range of technical journals. The IET is a multidisciplinary professional engineering institution that was formed from the Institution of Incorporated Engineers and the Institution of Electrical Engineers, which date back to the late 1800s in the United Kingdom. The IEC is a leading organisation that prepares and publishes international standards for electrical, electronic, and related technologies. The EIA is a trade organisation representing and sponsoring standards for the U.S. high technology community. The ECA is an associate of the EIA and manages EIA standards and technology activities for the electronics industry sector comprising manufacturers and suppliers.

The OMG is an international, not-for-profit computer industry consortium with open membership, which has task forces that develop enterprise integration standards for software related technologies and industries. Being an open group, the OMG operates under a different business model than the previously mentioned organisations. All of OMG's standards are published and accessible without fees on an open website. Unlike professional societies such as ISO, the IEEE, or INCOSE that raise operating funds through fees for the distribution of standards and journals, the OMG raises operating funds by other mechanisms.

The relationship between the systems engineering standards of ISO, the IEC, IEEE, and EIA can be understood by considering the level of detail provided and the breadth of scope across the system life cycle considered. The ISO/IEC/IEEE 15288 standard has the greatest breadth of scope but at a lesser level of detail. The IEEE 1220 standard provides a much greater level of detail but at a reduced breadth of scope. The EIA 632 and EIA/IS 632 standards are in between these ISO/IEC 15288 and IEEE 1220 standards in terms of level of detail and breadth of scope.

3.2   DEFINITIONS OF KEY TERMS

This part of the chapter introduces a half dozen foundational terms and definitions suitable for the model-based practice of System and Architecture Definition that will be introduced in Chapter 4 and demonstrated in the Part II tutorial case study. This core lexicon is not simply a list of terms and definitions. The succinct comparative and critical analysis of the definitions of the terms should provide the reader and student of the book a basic vocabulary and understanding to speak the language of architecture and systems knowledgably.

Commonly accepted definitions will be cited when available but when there is no accepted definition or if there are choices to be made between terms, definitions have been specified or modified in a way that provides a collection of terms that supports a model-based approach to architecture and systems engineering. This gives the collection conceptual integrity. It is appropriate to recall the Principle of Definition originally stated by F. P. Brooks (1995) that was adopted by Dickerson and Mavris (2010) and adapted by Dickerson et al. (2021):

> *Formal definition of concepts is needed for precision;*
> *prose definition for comprehensibility.*

This principle applies to all the fundamental terms of architecture and systems engineering which are defined in this book in terms of natural language because they admit interpretations into the graphical models of Unified Modeling Language (UML) and Systems Modeling Language (SysML). Although these two modelling languages are not formal, the underlying way in which they are used in the book is. Relevant concepts, methods, and applications have been published by the author over the past decade; the foundational research is published by Dickerson (2008; 2013), and by Dickerson and Mavris (2013). The summative research of the authors and their collaborators (Dickerson et al. 2021) offers a critical analysis of the definitions *architecture* and *system* as currently used in the relevant international standards.

A brief summary and critical analysis of foundational terminology for systems and architecture is now offered. A further lexicon of terms is presented in Annex A-3. At the simplest level of abstraction, the concepts of systems and architecture can be organised around those of combinations, partitions, and arrangements. These terms provide a simple intuitive description of a selection of elements joined into a whole, separation of the elements, and the composition of relations of the elements. The following foundational definitions will be used in the essential technical processes and structured methods in Chapter 4, and in the case studies in Sections II and IV:

    Structure
    Architecture
    Model
    Functionality
    System
    Engineering

Definitions of each of these terms follow below in the order above. The 'essential definitions' offered in this chapter are taken from recent summative research (Dickerson et al. 2021). They are subsistent in the sense of being an economical choice of generally understandable words that have a meaningful mathematical interpretation. The essential definitions are intended to complement the definitions from standards and authoritative sources.

3.2.1 Combinations, Partitions, and Arrangements

A *combination* is a collection of two or more elements selected or identified from a larger set that are in some sense joined together, i.e., combined. For example, a water molecule is a combination of two hydrogen atoms and one oxygen atom. However, no reference is made to relationships between the elements.

A *partition* of a set is a 'covering' of the set by a collection of subsets whose members are pairwise disjoint (i.e., have empty intersection). In mathematics, a covering of a set is a collection of subsets whose union is the set. The collection $\{\{a\}, \{b\}, \{c, d\}\}$ is a partition of the set $\{a, b, c, d\}$. A simple example from mathematics is the set of even and odd counting numbers which partition the counting numbers into two disjoint sets.

The term *arrangement* implies a composition of relations of selected elements of a set for a purpose, e.g., an arrangement of furniture in a room or making arrangements for a meeting. In music, the term can refer to a form of musical composition. The term can also have a specialised meaning such as an ordering or permutation in mathematics. However, ordering arrangements do not need to be linear and generally are not. A *scientific arrangement* or method may be defined as the gathering of individual objects into a synthetic whole for an experiment. This concept is very close to that of a system.

3.2.2 STRUCTURE

There are many definitions of the term structure and sometimes structure is confused with arrangement. In a dictionary definition, it is clear that as both a noun and an adjective this term has a stronger meaning than the term 'arrangement':

Noun: *The mutual relation of the constituent parts or elements of a whole as determining its peculiar nature or character; make, frame.*
Adjective: *Organized or arranged so as to produce a desired result. Also, loosely, formal, organized, not haphazard.*

Structure in the context of systems is distinguished from 'arrangement' in that it refers not just to the system elements but also to system properties.

It is worth noting that structure and organisation have been widely used in engineering but are generally not defined in the standards relevant to architecture and systems. Some of the definitions and usages of the term structure in engineering tend to focus on joining and connection of parts, e.g., in the ontology proposed by Sowa in *Knowledge Representation* (2000). As will be noted for the OMG definition of system architecture (in Section 3.2.4), the parts of a system may need to be isolated as well as connected. The essential definition put forth by Dickerson et al. (2021), on the other hand, considers both joining and separation:

Structure is junction and separation of the objects of a collection defined by a property of the collection or its objects.

This definition is at a higher level of abstraction than currently used in the standards but is readily applicable to physical examples. In civil engineering for example, a building is a collection of objects that includes rooms, which are joined as well as separated to achieve a defined purpose. Buildings in civil engineering are referred to as structures. The essential definition also has a mathematical interpretation, e.g., the property of divisibility by two defines a partition in the counting numbers that separates the even and odd numbers into disjoint subsets. This concept of structure therefore adheres to the Principle of Definition.

3.2.3 System Architecture

There are various engineering definitions of this term and the unscoped term 'architecture' is reported to have at least 130 definitions in various ISO standards. Three key definitions of architecture will be considered, one from the ISO//IEC/IEEE standards, another from OMG, and the other being the essential definition introduced by Dickerson et al. (2021). The definition used in ISO/IEC/IEEE 15288:2015 (ISO 2015) has been adopted from ISO/IEC/IEEE 42010:2011 (ISO 2011):

> *<System> architecture is the fundamental concepts or properties of a system in its environment embodied in its elements, relationships, and in the principles of its design and evolution.*

This definition is intended to complement the definition of *system* in the standard (ISO 2015). It lends itself to consideration of how the elements of a system are organised. However, the definition has certain ambiguities that make it difficult at best to apply rigorously to model-based methods. For example, the phrase "embodied in its elements, relationships" has led to debates as to whether architecture is a model. See the summative research (Dickerson et al. 2021) for further details.

3.2.4 Architecture: OMG Definition in MDA 1.0

Amongst the myriad definitions of system architecture is the one from the OMG offered two decades ago (OMG 2003):

> *The architecture of a system is a specification of the parts and connectors of the system and the rules for the interactions of the parts using the connectors.*

The definition is quite practical for understanding the software and physical implementation of a system architecture. However, it is silent on the *separation of the parts*. This is a problem for the civil engineering of a building or the systems engineering of a radar. The system architecture in both cases needs to specify how the elements are 'isolated' from each other. In a building, this could be sound isolation between rooms. In a radar, this could be high-voltage isolation between the transmitter and receiver chains. Thus, the definition could be improved by replacing the term 'connectors' with 'connectors and isolators'.

3.2.5 Architecture: Essential Definition

The essential definition of architecture is an abstract definition that is more precise and broadly applicable than the previous two. It is independent of any definition of system but can be used to express the relation between Architecture and accepted concepts of System currently used in the international standards. For example, system architecture can be understood as the fundamental form (or structure) of a system. The definition offered by Dickerson et al. (2021) expresses an intimate relation between architecture and structure:

> *Architecture is structural type in conjunction with consistent properties that can be implemented in a class of structure of that type.*

Such properties are said to be compatible with the type and are referred to as *architectural properties*. This language is unavoidably abstract because architecture is an abstraction, e.g., of a building or a system which is concerned with its structure and properties. For example, reliability properties can be implemented in a class of structures of parallel type for certain systems; a simple example being a string of lights (see also Chapter 12). The term *type* is a specialised property that can also be referred to as a *classifier*. It identifies the fundamental structure that is used to express the form of a system. An architecture is then a class of structure that is further defined by architectural properties. It can be thought of as a coupled pairing of properties: (*Structural Type, Architectural Properties*). This definition will provide a basis for specifying system models and their transformations with the precision needed for the rigorous process in Chapter 4.

3.2.6 MODEL

A general audience would understand the term 'model' as a *representation of something* or an excellent example according to a dictionary definition. Familiar examples are scale models, for example of an airplane or an automobile. Children's toys can also be models. Scientists and engineers have used models extensively, both physical and abstract, to understand and reason about real-world systems. In the twenty-first century, models have become increasingly digital even to the point of claims of models as 'digital twins' of physical entities. As excellent as a model might be though, by its very nature it is *not* the 'real thing'. Models are representations of 'referents', i.e., physical or logical entities.

The term model always *refers* to something, i.e., it is a *model of a referent*. Confusion can arise when a model is regarded as *The Model of the referent*. Amongst the anecdotal sayings about modelling and simulation is one that states, "all models are wrong, but some are useful". This issue is explained using simple mathematical models in biological science by Rosen (1993).

In order to avoid confusion between models and architecture, the well-established Tarski definition of 'model' from mathematics and logic is used throughout this book. This was explained in Chapter 2. It can also be used to complement natural language definitions. In mathematics, diagrams or symbolic expressions are generally used to represent models; but formally, models are *interpretations of sentences into structures*. The sentences can be regarded as concepts about the referent.

Tarski model theory (1954; 1955) offers the following simple but formal definition:

> *A model is a relational structure for which the interpretation of a sentence in the Predicate Calculus becomes valid (true).*

This is called a *fully interpreted first-order model; and is the basis for so-called model theoretic truth*. A relational structure in set theory is a set M and a collection (i.e., a set) of relations $\{R_a\}$ on M. In modern mathematics, concepts are realised in sets.

The term *concept* was discussed in Chapter 2. In precise terms, a concept is expressed as a *sentence* that has a *referent*, something that exists physically or abstractly. The terms concept, idea, and thought are used loosely both in general and in various knowledge domains. However, in modern psychology, *thought* is an activity of the mind and its existence can be identified by measurement of electrical activity of the brain. The term *idea* can be regarded as a thought that is suggestive of something, e.g., a purpose or course of action. In philosophy, a *concept* is a thought or idea which corresponds to a distinct entity or class of entities, or to its essential features. In Latin, the term *conceptum* means 'something conceived'.

When a sentence is expressed in the Predicate Calculus, the interpretation into a relational structure is exact. When a sentence is expressed in a natural language, the interpretation will necessarily be subject to the ambiguities of the language. Thus, concepts and terminology have an intimate relation that is central to any practice of model-based systems engineering.

3.2.7 FUNCTIONALITY

A straightforward adaptation of this term taken from dictionary definitions is:

> *Functionality is the purpose a system is intended to fulfil.*

This can be expressed by verb-noun phrases, such as 'track aircraft'. This construct is an implementation free view of a system. Functionality can be used to support a technical process of system definition and the functions within a system. Concept realisation is accomplished in part by system functions.

3.2.8 ENGINEERED SYSTEM

The most widely accepted definition of an (engineered) *system* is given in ISO/IEC/IEEE 15288: 2015 (ISO 2015):

> *a combination of interacting elements organised to achieve one or more stated purposes.*

Automobiles are engineered systems; computers are too. These are just two examples from everyday life. Almost every product or service used in everyday life can be viewed as a system using this definition. A careful examination of the definition shows that it is somewhat biased towards a 'white box' viewpoint of the system, i.e., the system elements. Nonetheless, it has sufficient precision to be applied to model-based processes and will be in Part II.

3.2.9 SYSTEM: ESSENTIAL DEFINITION

The essential definition put forth by Dickerson et al. (2021) is at a higher level of abstraction than the above definition. It can therefore be applied more broadly and also to systems that are not engineered.

> *A system is a set of interrelated elements that comprise a whole, together with an environment.*

A system then is not just a set or combination of elements; rather it is a pairing that is a coupling of two sets of elements: (S, E), where S is a set of elements defined as the system elements and E is a set of elements defined as the environment of the system. This definition will provide a basis for specifying system models and their transformations with the precision needed for the rigorous process in Chapter 4. The more intuitive but less formal term *form* can be used in place of *comprise*. Thus, the system elements can then be said to *form a whole*.

The term System is widely used in science, mathematics, and engineering. A general system can be regarded as a combination of interrelated parts that have properties as a *whole* not possessed by any of the parts alone. Our solar system is a natural system, i.e., an example of a general system occurring in nature. Other physical entities from atoms to animals to the universe itself are systems too; so too are weather systems. The list goes on: business systems and enterprises, economic and political systems, and so forth. It is perhaps only limited by the imagination. The term *interrelated* is more general than and is inclusive of the term *interacting* that is used in the current standards, i.e., interaction can be a type of interrelationship. Abstract systems such as the counting numbers also conform to the essential definition but not to the definition used in standards. Engineered systems are distinguished from other systems by human intent.

3.2.10 ENGINEERING

The nature of modern engineering is based on its evolution in Western civilisation (Finch 1951). Engineering of a product begins with craft but must be honed using the laws of science for precision and repeatability in commercial production. The following definition (Dickerson et al. 2021) aligns well with the essential definitions of structure, architecture, and system that have been proposed:

> *Engineering is a practice of concept realisation in which relations between structure and functionality are modelled using the laws of science for the purpose of solving a problem or exploiting an opportunity.*

It is worth noting the architectural viewpoint in this definition. The association of structure and property (i.e., functionality) conforms to the pairing in the essential definition of architecture.

3.3 MATHEMATICAL INTERPRETATION OF ARCHITECTURE

As noted in the definition of partition (Section 3.2.1), the set of even and odd counting numbers is a partition of the counting numbers into two disjoint sets. The essential definition of structure conforms to this simple mathematical example. The property of 'divisibility of a counting number by the number two' defines the partition. Therefore, the essential definition of structure adheres to the Principle of Definition because it has a meaningful mathematical interpretation. Using an argument similar to the structuring of the counting numbers into disjoint subsets, the essential definition of architecture can also be demonstrated to adhere to the Principle of Definition by specifying a structural type paired with an architectural property. A straightforward example with engineering utility is a *constraint architecture* that defines a feasible region of a design space. The properties of interest are the constraints on the space. The structure will not be as simple as that of a partition. Rather than separating subsets the constraint architecture will be concerned with joining them through set intersection. This example (as well as Chapter 12) is especially useful not just because of its engineering utility but also because it can be used to demonstrate why not all properties can be implemented in a given class of structures.

3.3.1 Application to Engineering the ADAS ECG

The constrained operating space of the diesel engine in Figure 2.1 for the emissions control governor is a typical example. The state variables were each restricted to an interval of values that depended on an initial state. For a single real-valued variable, one type of a such a constraint is in the form $a \leq x \leq b$, where a and b are distinct. If $a = b$ then the constraint is degenerate and not really an interval. The numbers a, b are referred to as the bounds of the interval, or more loosely as constraints; but mathematically the constraint is a relation between the bounds and the variable. Another type of constraint is in the form $a < x < b$, which is an open interval. In this case, if $a = b$, the interval is the empty set and the constraint is void (and invalid). These two types (which involve only a single variable) provide the simplest examples of a constraint structure. They will be used both to demonstrate a simple mathematical interpretation of the essential definition of architecture and to gain insight into its careful arrangement of wording.

For the case of a single real-valued variable, a constraint is then a type of relational property that defines a collection of intervals. There are four types that define intervals in one of the following four forms:

     i. Closed intervals: $[a, b]$
     ii. Open intervals: (a, b)
    iii. Semi-open intervals
         a.   Upper semi-open: $[a, b)$
         b.   Lower semi-open: $(a, b]$

The corresponding classes of constraint structures will be denoted as I_C, I_O, I_U, and I_L. In engineering practice, the structure with open intervals represents strict constraints. The upper semi-open intervals are well-suited for the constrained operating space of the diesel engine in Figure 2.1. The inclusion of the lower bound in the intervals would correspond to the initial state that a cruise or acceleration starts from. In general, the semi-open intervals can be useful for constraints on time in dynamic systems. The inclusion of the lower bound would correspond to the start time of a process and the strict open upper constraint is a requirement to finish *before* a specified end time. Similarly, the inclusion of the upper bound in a lower semi-open constraint could signal that the process must finish exactly at the end time.

Only one architectural property will be needed to define a class of constraint architecture. This is jointness (expressed through set intersection). This is how feasible regions are formed in a design space. The mathematical statement of jointness is that the non-empty intersection of any two intervals in the specified structure is also a member of the structure. This means that every pair of constraints that are consistent (have non-empty intersection) can be replaced by a single constraint of the same form. In the above classification scheme of structures, only three of the four classes of structure satisfy this property. The class I_C clearly does not. Consider two constraint intervals: $[a, b]$ and $[c, d]$. The joint constraint is defined by intersection of the two intervals. Consider the case of $c \leq b$; the intersection is non-empty and would result in the interval $[c, b]$. If $b = c$, the intersection degenerates into a point and is not an interval.

Thus, I_O, I_U, and I_L are suitable for specifying joint constraints but I_C is not. The architect then has three classes of structure to choose from for the purpose of specifying a *constraint architecture* for the design space and recommending it to the engineering team. The final choice can be made based on other criteria. For the constrained operating space of the diesel engine in Figure 2.1, the statement 'without violating the emissions constraints' in the textual descriptions associated with the joint constraints would correspond to the class I_L. Violating a constraint $x = b$ in the operating space means $x > b$. Thus, the operating interval would be in the form $(a, b]$. However, based on the architectural analysis of the operating space, it might be more useful to choose the class I_U so as to use the initial states more effectively.

3.3.2 Correct Use of Mathematical Language

The use of language in the application of architecture to the constrained operating space is correct and precise. However, as with any practice of engineering, many underlying details have been suppressed so as to make clear the message to the intended audience of the discussion. A skilled architect should be able to speak and write 'bi-lingually' in this way. In what follows are the details of the correct use of mathematical language that underly the constraint architecture example for the operating space of the diesel engine.

First, it is important to understand the relation of the mathematical terms class and type, and to grasp the distinction. Annex A-1 provides a distilled but comprehensive background on this and other essential mathematics that underly the concepts presented in this book. In general terms, a class is a collection of (mathematical) objects that is defined by a property. The class can be said to implement the property, but it is not the property. In the separation of the counting numbers into even and odds numbers, the property that defined the partition was 'divisibility by the number two'. Even and odd numbers are classes of counting numbers that are realised (instantiated) in a set of counting numbers. In a constraint structure, the property is the constraint type. For a single real-valued variable when the pairs of distinct bounds (a, b) are varied over all values in which $a < b$, the result is a collection of intervals that are subsets of the set of all real numbers. The collection is a relational structure (as introduced in Chapter 2, Section 2.3).

When properties are used for classification schemes they are often referred to as classifiers. The term *type* is a property that is generally used in this sense. Thus, the even numbers are a type of counting number. This can be represented symbolically as «even», i.e., the name of the property that defines the set of even numbers as a subset of the counting numbers. This is referred to as a *type specification*. Constraint structures in the example belong to a class of structures defined by the type «unary relational». This level of formality is needed to formally explain that not all properties can be implemented in a given class of structures.

The associated type specifications can be denoted as «closed unary relational», «open unary relational», «upper unary relational», and «lower unary relational». In the language of the essential definition of architecture, these are four structural types that can be paired with one or more architectural property to define *constraint architecture*. When the property is consistent with the structural type it can be implemented in a class of structures defined by the type and becomes *a class of constraint architectures*. More simply it will be referred to as *a constraint architecture* when it has been realised in a set. In the example above for the constrained operating space of the diesel engine, only three of the four structural types were consistent with the architectural property of joint constraints.

As discussed in Chapter 2, Section 2.3, in the practice of science and engineering this formality is normally suppressed but if the models and concepts do not conform to this formal basis, they will be logically incorrect and exhibit errors. The reader is challenged to interpret the ISO and OMG definitions of system architecture (provided in Sections 3.2.3 and 3.2.4) into this straightforward example or in any other mathematically meaningful way. On the other hand, the mathematical interpretation of the essential definition (provided in Section 3.2.5) is seen to be clear, precise, and indeed powerful because of the breadth of interpretations into real engineering problems that it can support.

3.4 LOGICAL MODELS OF CONCEPTS

The purpose of this part of the chapter is to show how natural language definitions of terminology can be modelled with the same level of formality as the science-based example in the Chapter 2. The ISO/IEC/IEEE 15288:2015 definition of system will be used as an example. The system and architecture definition processes specified in Chapter 4 require precise yet intuitive definitions of the two terms system and architecture that have been normalised against each other. The logical model of system presented in this chapter will be evolved using the essential definitions. The result will be an integrated model of the concepts of system and architecture that is suitable for the specification of mathematically based technical processes and structured methods that implement the processes.

Concepts expressed in natural language can also be interpreted into sentences in the predicate calculus and then into a model. The process of doing so exposes the inconsistencies and ambiguities that occur in the common usage of natural language. This is an example of the Principle of Definition. This type of model, like any other model, may not be unique but it serves as a point of reference for analysis and discourse that should lead to an agreed upon model. The correct use of a formal language in this way is beyond the grasp of all but perhaps a few experts.

Early progress towards making such a process of expression more intuitive was made by Peter Chen (1976) in the 1970s using a graphical language, Entity–Relationship (E–R) Diagrams. These are still in use today for relational data bases. Around this same time period, John Sowa (1983) developed a similar type of diagram that was specialised to the predicate calculus. These were called Conceptual Graphs, which he used to develop a framework of Conceptual Structures. As presented by Sowa, this is essentially a process for representing knowledge inside computer systems; but it is more general and can be applied more widely. In his framework, a concept is a mental interpretation of a percept (an impression of the physical world or a thought about an abstraction). Basically, any thought about something (a referent) that is capable of being expressed in natural language can be regarded as a concept in this more formal sense. Conceptual Graphs have been standardised as part of information technology for many years. The most recent version is ISO/IEC 24707:2018 (ISO 2018).

By the late 1990s, the notations of E–R Diagrams and Conceptual Graphs had been absorbed into the UML that has been a standard from the Object Management Group (OMG) for object-oriented software development for over two decades. It can also be used to represent sentences in the predicate calculus in a way that is intuitive and graphical. Entities are represented as classes, which in the graphical notation of UML are rectangles with a name (typically a noun) that identifies the class. A class is a collection of objects of the same type. These can correspond to the variables in a formal predicate. The relation between the variables, which in mathematical logic is represented by a predicate letter, is resolved into relationships between the variables which are represented by lines between the classes and typically annotated with a verb or verb phrase that expresses the meaning (semantics) of the relation.

A concept expressed in natural language that has been interpreted in terms of a sentence in the predicate calculus will be referred to as a *logical model of the sentence*. The objective of the logical modelling of a sentence is to extract the relations comprising the sentence using a formal language to derive a minimal model that is complete and captures the intended meaning of the sentence. Key words are selected that remain undefined in order to avoid the circularities possible in natural language. This approach follows the axiomatic approach of mathematics. See the foundational work (Dickerson 2008) that linked the scientific concept of a mathematical model in Hamiltonian mechanics with the concept of model in mathematical logic using a procedure for the logical modelling of sentences. This procedure was demonstrated in detail in the first edition of this book (Dickerson and Mavris 2010) and is provided in Annex A-2.

The outcome of the procedure will now be illustrated using the term system. The definition currently adopted in the most widely accepted standard for systems is (ISO 2015) "a combination of interacting elements organized to achieve one or more stated purposes". Although this particular standard has not yet adopted the use of logical models of sentences, numerous others have, e.g., ISO/IEC/IEEE 42010:2011 (ISO 2011) to name just one. The diagram depicted in Figure 3.1 of this particular definition is organised around six key words, only one of which is a verb (achieves).

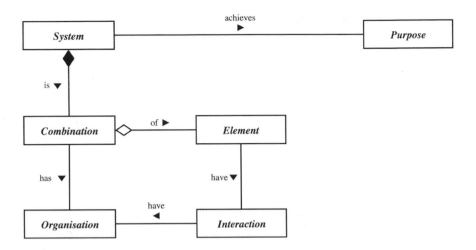

FIGURE 3.1   Logical model of a standards-based definition of system

Two of the adjectives have been converted to nouns and are represented as properties possessed by two of the nouns occurring in the defining sentence. The third adjective (stated) has been suppressed and would be best represented as a verb (state) but there is no suitable noun in the sentence to serve as the subject, e.g., 'Stakeholders state'. Altogether, six relationships form a predicate that has been expressed by the natural language sentence. As a formal sentence in logic, the overarching predicate statement would be: $v = P(u_1, u_2, u_3, u_4, u_5)$ where v takes on the value System, and u_1, u_2, u_3, u_4, u_5 take on the values Purpose, Combination, Element, Interaction, and Organisation, respectively. In this formal sentence the equal sign ($=$) is interpreted as 'is'. The statement is a well-formed formula in which every variable has been assigned a value, i.e., the formula has been quantified. It therefore is a sentence in the predicate calculus.

The predicate letter P can be resolved into five further predicates of two variables each. For example, $P_1(v, u_1)$ could be the predicate 'System achieves Purpose'. The resulting formula would be

$$[P_1(v, u_1)] \wedge [v = u_2] \wedge [P_2(u_2, u_3)] \wedge [P_3(u_3, u_4)] \wedge [P_4(u_2, u_5)] \wedge [P_5(u_4, u_5)].$$

In this predicate formula, the symbol $\wedge$ is the 'logical and'. Thus, all six relationships in the diagram are represented in the predicate formula in terms of a conjunction ('logical and' statements). In practice, this level of technical detail would be suppressed. However, the point is that the graph in Figure 3.1 is a structure in which a sentence in the predicate calculus has been interpreted. The UML diagram in the figure therefore represents a first-order model in the mathematical sense and has the precision and consistency needed for mathematical analysis. The UML notation in the diagram also has symbols for aggregation and composites that can be useful for modelling systems. These will be explained in Chapter 8.

3.5 MODEL-BASED SYSTEMS ENGINEERING

Now that an accepted definition of system has been logically modelled and a definition of engineering has been proposed, it would be appropriate to reflect upon an appropriate definition of the term systems engineering. One approach would be to use the technique of viewpoints and views that are commonly practiced in engineering. The result from doing this will be compared with an accepted statement of model-based systems engineering. Recall that a viewpoint is a technique for abstraction to focus on particular concerns regarding a concept. The term abstraction as used here means the process of suppressing selected detail to establish a simplified model. A viewpoint can be thought of as a perspective. A view, on the other hand, is a viewpoint model, a representation of the concept from the perspective of a particular viewpoint.

As noted in the first edition of this book (Dickerson and Mavris 2010), it is very natural to then consider that

Systems engineering is the practice of engineering from a systems viewpoint.

This definition could be used with any definition of engineering. It also has a syntax that is consistent with the terminology of other engineering disciplines, for example:

- Chemical engineering could be considered as the practice of engineering from the viewpoint of chemistry.
- Structural engineering could be considered as the practice of engineering from a structural viewpoint.

This syntax also makes clear the difference in roles between systems engineers and other engineers in the multi-disciplinary environment in industry.

As demonstrated in the logical model in Figure 3.1, the term system can be expressed as a well-defined concept. Therefore, it would also be natural to replace the term concept in definition of engineering presented earlier in this chapter with the term system:

Systems engineering is a practice of system realisation in which relations between structure and functionality are modelled using the laws of science for the purpose of solving a problem or exploiting an opportunity.

This definition is also consistent with the concept of *engineering of systems*. All three of these concepts are model-based by the nature of the definition of the term engineering given earlier in this chapter.

The 2007 INCOSE Vision 2020 as noted in the INCOSE Systems Engineering Handbook defined model-based systems engineering as, "the formalised application of modelling to support system requirements, design, analysis, verification, and validation activities beginning in the conceptual design phase and continuing throughout development and later life cycle phases" (INCOSE 2015). The scientific basis of engineering expressed in this chapter together with the formal first-order model theory that has been demonstrated for the concepts of system in this chapter clearly establishes the approach taken in this book as a qualified approach to model-based systems engineering, especially when these concepts are used to implement the technical processes of accepted standards such ISO/IEC/IEEE 15288:2015 (ISO 2015). This implementation will begin at a foundational level in the next chapter and be carried forward with increasing detail throughout the book.

4 Structured Methods

KEY CONCEPTS

Separation of concerns
Association and cohesion
Model transformation
Synthesis of models

The basic concepts of Structured Analysis and Design were reviewed in the brief histories of architecture and model-based systems engineering in Chapter 1. Initially introduced by Yourdon, these concepts have evolved and become the basis of standards and practices in the twenty-first century such as the OMG Model Driven Architecture for software development. The *principle of separation of concerns* is fundamental to these concepts and is broadly applicable to engineering and systems. Architecture, analysis, and transformations between models are central to a structured approach to system design. The logical foundation of rigorous methods for model specification and transformation used throughout this book is based on a synthesis of the concepts introduced in Chapter 2 with those of Yourdon. This is the basis of a relational framework for Structured Analysis and Design.

A system realisation of the concept for an emissions control governor (ECG) investigated in Chapter 2 will be used in this chapter to illustrate how the specification and transformation of first-order models can be used for precise interpretation of a science-based solution into a model-based specification of an information-intensive systems solution. This will complement and illustrate the foundational discourse of how systems, architecture, modelling, and engineering can be bound together in a straight-forward intuitive way that is rigorous and practical. The remainder of the book will be concerned with development and explanation of the requisite technical details for application to more complicated problems than the simple examples in this foundational section.

4.1 SCOPE OF CONCERNS AND SYNTHESIS OF CONCEPTS

Two core technical processes specified in ISO/IEC/IEEE 15288:2015 (ISO 2015) are explored in depth in this book: System Requirements Definition and Architecture Definition. Altogether there are three requirements technical processes in the ISO standard. The other two are Business or Mission Analysis, and Stakeholder Needs and Requirements. However, an in-depth exploration of all three is beyond the scope of the book and indeed could constitute a separate book. In the tutorial and case study materials, the outputs from the other two processes for requirements are assumed

DOI: 10.1201/9781003213635-4

(given as starting points). Core elements of the Unified Modeling Language (UML) that are shared with the Systems Modeling Language (SysML) will be used for specifying essential system models and architecture. Limiting the initial specification of models to UML serves two important purposes. First, as seen in Chapter 2, is that its form is close to that of mathematical logic and set theory. Simple guidelines and procedures can therefore be given that make system models and architecture more rigorous and traceable because they reflect an underlying expression in the predicate calculus. The other purpose is that the core models specified in UML can be given directly to software engineers for code development. The same models can be extended or elaborated into SysML for use by other engineers for design and development. If maintained properly, these 'essential models' can be used to maintain the conceptual integrity of the system through a comprehensive architecture that enables the design and development of hardware and software in a holistic concordant way.

4.1.1 Alignment of Definition Processes to System Life Cycle

The 14 technical processes specified in ISO/IEC/IEEE 15288:2015 are easily aligned to the traditional Systems Engineering Vee as in Figure 4.1. This arrangement lends itself to a 'top-down' approach that progresses from the top left of the Vee then down the left side using the traditional method of system definition and decomposition. Implementation of the system elements of the design is at the bottom of the Vee. Integration and Verification then progress in an iterative loop until prototypes and the finished system are ready to be validated. After validation is accomplished, the system is transitioned to the customer where the life cycle is completed (deployment through disposal).

This process was originally developed in the mid to late 1990s in the context of large-scale systems in the defence and aerospace industry that had long development cycles. Thus, the linear progression through the Vee was not a significant factor in system development. However, as systems engineering expanded from these origins into the commercial domain, this has become an issue. This is especially true for software development. In the model-based approach to architecture and systems

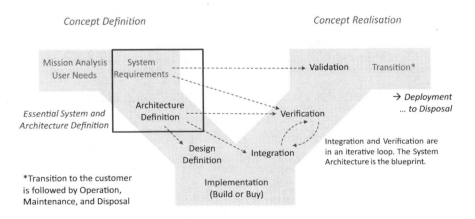

FIGURE 4.1   Alignment of Essential Definition processes

engineering presented in this book, the process of system definition and decomposition will be subsumed into one of model specification and transformation. This permits both vertical and horizontal reach from Architecture Definition throughout the Vee, not just to Design but also to Integration, Verification, and Validation. The outcome of the definition process should be a model-based specification of the system that is ready for design and might also be ready for implementation (i.e., sufficient detail for 'build or buy' of the system elements). Furthermore, in this approach a linear progression through the Vee need not be followed; thus avoiding many of the issues associated with applying it to complex systems and commercial development.

How this can be done in a commercial environment was a subject of a five-year research programme with an automotive original equipment manufacturer (OEM) that was jointly sponsored by the government and the OEM (Dickerson and Ji 2018). The research was concerned with virtual design and prototyping for increased speed to market. The long-standing practice of specifying configuration items (CIs) during Architecture Definition can also be employed to an advantage in the model-based approach. CIs are hardware, software, or composite elements at any level of the system hierarchy designated for formal control of form, fit, or function. They are characterised by being replaceable as an entity, having defined functionality, and a unique specification (INCOSE 2015). CIs can and should be specified as model elements in the system model. This provides a criterion for a minimal level of modelling because every CI is managed under formal control.

Defining a system as either a combination of interacting elements or simply a set of elements initiates a *System and Requirements Definition* process that begins with a simultaneous decomposition process that identifies the relevant elements. The definition process must then use this identification to transform the outcomes of the mission analysis and stakeholder needs processes into a technical view of the system. This includes

- System boundary, elements, functionality, and process specifications
- Functional, performance, and non-functional requirements; and constraints
- Traceability of system requirements to stakeholder requirements

The specifications, requirements, and constraints are system concepts.

Architecture Definition is concerned with how stakeholder and system requirements relate to system structure. The salient features are:

- Defining structures associated with a system to implement the system concepts
- Stating the concepts as properties to be implemented in the structures
- Interpreting the statements into the structures (i.e., specifying models)
- Defining relations amongst the structures to include transformation and synthesis
- Normalisation of concepts, semantics, and relations across the models
- The structural relations include interfaces between the processes of the system elements

A process of model specification and transformation then follows from the interpretation of the concepts into the structures and the relations amongst the structures.

In order to use an architecture defined by this process to enable Design Definition and implementation specification, it must be further refined to define how:

- Structures associated with the system response under intended and alternative conditions
- Control structures can be defined for precise implementation of responses

The outcome of the above definition and refinement processes is a collection of models that are bound by a system architecture that is robust, and are suitable for specifying implementation of the hardware, software, and composite elements of the system, or to enable an iterative design definition process that does. This collection of models bound by the architecture will therefore be referred to as an Implementation Model. A holistic comprehensive system architecture will include a software architecture that can be deployed to software developers. Deployment to state machines will define how structures associated with the system change state in response to the occurrences of events.

4.1.2 MODELLING LANGUAGES AND LOGIC

The organisation of diagram types in SysML and their foundation in UML is depicted in Figure 4.2. SysML is a general-purpose modelling language similar to the UML that is formally defined as a profile of UML 2.0. This means that the SysML metamodel (OMG 2017a), i.e., the syntax, semantics and modelling rules are built around the metamodel of UML (OMG 2017b) by means of extensions. These are in the form of stereotypes, tagged values and constraints. Specifically, SysML inherits a subset of the model elements defined in UML and extends other model elements to facilitate generic system modelling. The commonalities and differences between the two modelling languages can be better visualised at the level of diagram types rather

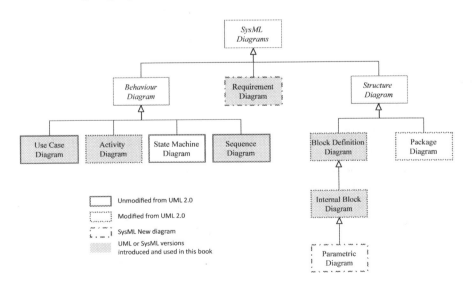

FIGURE 4.2 The Systems Modeling Language

than at the level of model elements. Adapted from SysML specification, Figure 4.2 provides this visualisation from the perspective of SysML diagrams. Note that the SysML Block Definition diagram and Internal Block diagram are analogous to the UML Class diagram and Composite Structure diagram (OMG 2017b) respectively with moderate modifications. Other well-known UML Profiles include, for example, the UML profile for the Modeling and Analysis of Real-time and Embedded Systems (often abbreviated by MARTE) (OMG 2019). This is a modelling language that extends UML specifically for application to embedded systems.

The essential definitions of structure, architecture, and system presented in Chapter 3 have a mathematical basis that is suitable for deriving technical processes for systems engineering (Dickerson et al. 2021). The underlying concepts of structure and behaviour are more general than those of the SysML diagram types of the same name. As noted earlier, the core diagrams of UML that are in common with SysML can be used to specify system models and architecture that are mathematically based. This is similar to how a definition of 'system' was modelled with UML Abstract Classes in Chapter 3, yielding a graphical representation of a logical predicate formula. Use Case diagrams are particularly straight forward to model with this level of rigor when some simple guidelines and procedures are adhered to. Because this type of diagram is the simplest and most ubiquitous in systems, software, and standards, it merits a detailed explanation.

A Use Case diagram only has four types of graphical elements: a figure (which represents an UML Actor), a line (which represents an UML Association), a rectangle (which depicts the system boundary), and an oval (which represents an UML Use Case). It is a widely accepted practice that each use case should be written as a verb-noun phrase. Recall from Chapter 3 that this type of phrase is a predicate. Thus, similar to the logical modelling of sentences, use cases can be written as predicates $P(v)$ that associate elements of the environment (actors) with the system as a black box entity. The association can be expressed as a sentence in one of two general forms: (i) System <interacts with Actor>, or (ii) Actor <interacts with System>. The logical formula for the sentence is $u \wedge P(v)$ where the predicate variables u, v each exclusively take on the value of either the actor or system name. (The symbol $\wedge$ is the 'logical and'.) The formula is indeed a sentence in the predicate calculus because the variables have been quantified (by the actor and system names). All of these concepts and notations are captured succinctly in Figure 4.3.

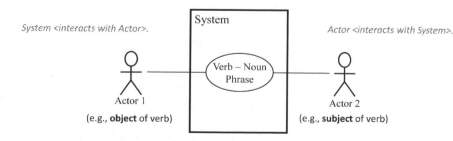

FIGURE 4.3   Relation of predicates to use cases

4.1.3   Functional Requirements for the ADAS ECG

As previously noted, the Use Case diagram is the most widely used diagram in UML and SysML. Functional requirements can be easily and precisely captured in this simple graphic when written as simple sentences using predicates. Any *system* realisation of the concept for the ECG investigated in Chapter 2 must include a functional requirements specification. The annotated diagram in Figure 4.4 fulfils this need. Because it follows the guidelines for the relation of predicates to use cases, it offers textual (prose) statements that can be represented formally in the predicate calculus but are presented in a simple intuitive graphical form that is easy to communicate to customers and engineers alike. This is precisely the intent of the Principle of Definition and gives the system architect a means to fulfil their role regarding conceptual integrity of the system; thus, adhering to two of the principles of architecture and systems engineering.

Complementary to the Use Case diagram in Figure 4.4 is the matrix representation in Table 4.1 that depicts the information content of the diagram. The specific interpretation of the symbols is as follows:

- Actors:
    - $A_1 = \text{ADAS}$
    - $A_2 = \text{Engine Emission}$
    - $A_3 = \text{Engine}$
- Predicates:
    - $P_1(v_1) = \text{Make Torque Demand}$
    - $P_2(v_2) = \text{Control Emissions}$
    - $P_3(v_3) = \text{Govern Torque Demand}$
    - $P_4(v_4) = \text{Govern Engine Speed}$
- Sentences: $W(A_i, S; u \wedge P(v))$
    - $A_i = i\text{th Actor}$
    - $S = \text{EGC System (black box)}$
- Logical Symbols: $\wedge$, $\vee$, $\neg$
- Quantification: u, v exclusively take on the value of A_i or S

Written in symbolic notation this representation can be machine readable. This sort of representation should be running in the background of any commercial modelling tool for a practice of engineering that uses UML and SysML.

When the traditional process of system definition and decomposition is subsumed by one of model specification and transformation, graphs and tables such as in Figure 4.4 and Table 4.1 can be interpreted as matrices that can be used for various types of analysis. For initial modelling of use cases the matrices are very simple. The matrix for the actors is an $m \times 1$ vector (in the table, $m = 3$) and the matrix for the use case predicates is a $1 \times n$ vector (in the table, $n = 4$). The $m \times n$ matrix depicts how the actors are associated with the use case predicates by the sentences. Because the sentences establish interrelationships between elements of the environment and

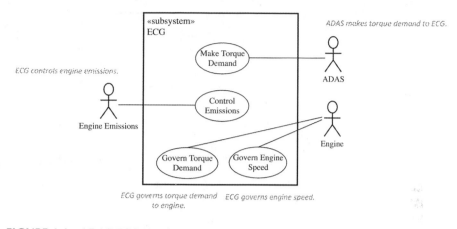

FIGURE 4.4 ADAS ECG Use Case diagram

TABLE 4.1

ADAS ECG Use Case Diagram in Matrix Form

Actors	Predicates	$P_1(v_1)$	$P_2(v_2)$	$P_3(v_3)$	$P_4(v_4)$
A_1		$W_{1,1}$	...	...	...
A_2		...	$W_{2,2}$	...	...
A_3		...	...	$W_{3,3}$	$W_{3,4}$

properties of the system (e.g., functionalities expressed through use cases), the model represented by a Use Case diagram is an elementary one that expresses the concept of system functionality as a set coupled with an environment via interrelationship.

In general, there can also be relations between the actors, as well as relations between the predicate variables. In this case, the one-dimensional vectors for the actors and predicates must be expanded into $m \times m$ and $n \times n$ matrices, respectively. This is depicted by the relational framework of matrices in Figure 4.5, which is formed by three matrices, M, Q, and N. Rather than listing the actors and predicates, as in Table 4.1, the variables of the predicates and variables associated with the actors are listed. In the design of systems, the variables are generally metric variables that quantify the requirements and the system level characteristics.

For the specification of system models, the M-matrix is an array of relationships between design objectives $z_1, z_2, \ldots z_m$ and the N-matrix is an array relationships between the system level characteristics $x_1, x_2, \ldots x_n$. The relationships in the cells of the matrices are the predicates. As with the use case table, the predicates in the two matrices are joined by sentences. These are in the $m \times n$ Q-matrix. The complexity of system design and specification now becomes evident. Furthermore,

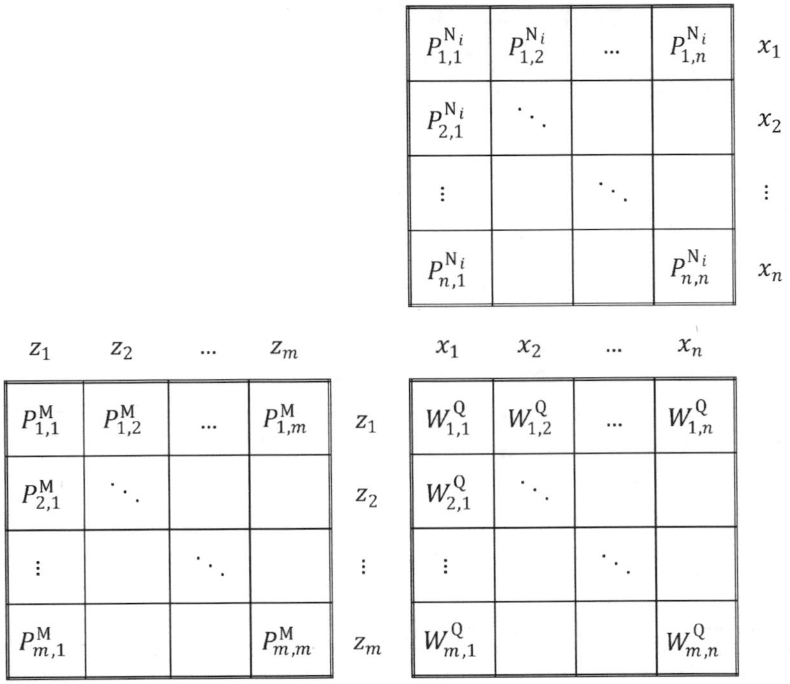

$$
\begin{array}{|c|c|c|c|}
\hline
P^{N_i}_{1,1} & P^{N_i}_{1,2} & \cdots & P^{N_i}_{1,n} \\
\hline
P^{N_i}_{2,1} & \ddots & & \\
\hline
\vdots & & \ddots & \\
\hline
P^{N_i}_{n,1} & & & P^{N_i}_{n,n} \\
\hline
\end{array}
\begin{array}{l}
x_1 \\ \\ x_2 \\ \\ \vdots \\ \\ x_n
\end{array}
$$

$$
\begin{array}{cccc}
z_1 & z_2 & \cdots & z_m
\end{array}
\qquad\qquad
\begin{array}{cccc}
x_1 & x_2 & \cdots & x_n
\end{array}
$$

$$
\begin{array}{|c|c|c|c|}
\hline
P^{M}_{1,1} & P^{M}_{1,2} & \cdots & P^{M}_{1,m} \\
\hline
P^{M}_{2,1} & \ddots & & \\
\hline
\vdots & & \ddots & \\
\hline
P^{M}_{m,1} & & & P^{M}_{m,m} \\
\hline
\end{array}
\begin{array}{l}
z_1 \\ z_2 \\ \vdots \\ z_m
\end{array}
\quad
\begin{array}{|c|c|c|c|}
\hline
W^{Q}_{1,1} & W^{Q}_{1,2} & \cdots & W^{Q}_{1,n} \\
\hline
W^{Q}_{2,1} & \ddots & & \\
\hline
\vdots & & \ddots & \\
\hline
W^{Q}_{m,1} & & & W^{Q}_{m,n} \\
\hline
\end{array}
\begin{array}{l}
z_1 \\ z_2 \\ \vdots \\ z_m
\end{array}
$$

FIGURE 4.5   Relational framework

the interrelational nature of systems adds a further complexity: the sentences in each row of the Q-matrix can convey relationships between the objectives into relationships between the system characteristics. These will be in conjunction with any preconceived relationships between the characteristics. Consequently, the interrelational nature of systems implies that there is a family of N-matrices that must be used for analysis and design of the system (i.e., determining the best values of the system characteristics to achieve the design objectives). This is why the superscript of the predicates in the N-matrix in Figure 4.5 is indexed by the rows of the M-matrix. This family of matrices $\{N_i\}$ is in fact a *relational structure* on the system design characteristics $\{x_1, x_2, \ldots x_n\}$. Thus, the first-order model theory introduced in Chapter 2 is not just an interesting insight. It is at the very core of architecture, analysis, and system design. Furthermore, the Q-matrix becomes a model transformation from the logical structure of the problem to be solved into the relational structure of the system.

4.2   STRUCTURED ANALYSIS

Chapter 1 introduced the Yourdon Structured Analysis method which is based on a concept of separation of concerns. The Principle of Structured Analysis can be summarised succinctly as follows:

The specification of the problem should be separated from that of the solution.

Two types of system models are used in structured analysis:

Essential Model (implementation-free representation)
Implementation Model

These two types of models are linked by interpretation of a behavioural model into one for implementation that is a realisation of the essential model. The Implementation Model is a specific representation of the hardware and software needed to realise the system.

The Essential Model is comprised of the Environmental Model and the Behavioural Model (Yourdon 1989). The Environmental Model defines the boundary between the system and the environment. It provides a short description of the system purpose, a context diagram, and an event list. The Behavioural Model identifies what the flows are through the system and establishes controls over the system flows. In this way, a model of what is flowing through the system is developed. It also determines the possible system states and transitions.

4.2.1 FRAMEWORK FOR STRUCTURED ANALYSIS

Structured Analysis can be implemented through a framework of model specification and transformation as depicted in Figure 4.6. This corresponds to the technical processes in the upper left corner of the Vee diagram in Figure 4.1 and proceeds in two steps. Based on the essential definitions, the system is formed of two sets of elements: the system elements and the elements of the environment. The System Requirements Definition process from ISO/IEC/IEEE 15288:2015 (ISO 2015) can be applied to transform a specification of the environment based on mission analyses and user needs into a technical view of system functionality. If the resulting technical view of the system has sufficient detail to define the functionality needed for all users of the system, then this step in the structured analysis process is finished (subject of course to agreement between the stakeholders). Otherwise, further elaboration of functionality is needed. This can entail identifying and defining new functionalities not previously discovered in the earlier Requirements Definition processes or applying system decomposition to specify lower level functionalities. The output of this first step in the structured analysis process is the Environmental Model. Figure 4.4 provides an example for the ADAS ECG.

The second step, as indicated in Figure 4.6, is to transform the system functionality into system behaviour (which shows the internal processes). This process begins when system functionality has been defined in sufficient detail to enable specification of system activities and basic flow of actions that enable each functionality. When the basic flows have been specified to a sufficient level of detail that exhibits how the system processes will deliver the intended functionality in the intended environment, the behaviour modelling is complete. The fourth part of this chapter provides an example of basic flow of actions for the ADAS ECG. The specification of the Behavioural Model completes the specification of the Essential Model. The structured analysis process should be applied to the system elements in the same way as it is to the (overall) system.

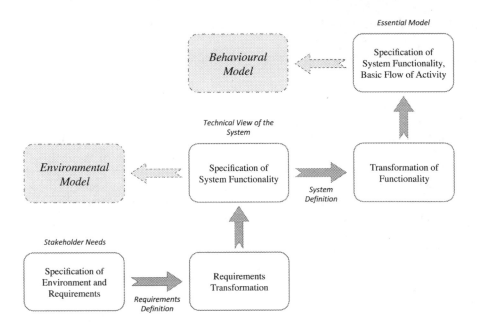

FIGURE 4.6   Framework for Structured Analysis

4.2.2 MODELS FOR STRUCTURED ANALYSIS

The Requirements and System Definition processes can be implemented by model specifications and transformation to create the Environmental and Behavioural Models. For the Environmental Model, the Use Case diagram provides a graphical model that defines the boundary between system and environment, and provides a description of the system functionality (purpose). The transformations are: Use Cases → Included Use Cases (functional decomposition) → Use Case descriptions. For the Behavioural Model, the Activity diagram provides a graphical model that identifies what the flows are through the system and will establish controls over the system flows. Thus, two classes of structure are realised in the Essential Model: a partition (of the actors and system in the use cases) and a basic flow structure (in the Activity diagram).

4.3   STRUCTURED DESIGN

The model-based approach for software development from a behavioural viewpoint introduced by Yourdon (1989) is an approach to implementing the Essential Model of the system:

> *Systems should be comprised of modules each of which is highly cohesive but collectively are loosely coupled.*

This succinctly states the Principle of Structured Design. Cohesion minimises interactions between elements not in the same module, thus minimising the number of connections and amount of coupling between modules. The strongest form of cohesion in the Yourdon approach was based on functionality. His ranking was only suggestive and not based on any sort of rigorous analysis. The following is a slight re-ordering of Yourdon list with some explanation:

Coincidental	*concurrence without apparent cause*
Logical	*e.g., association by conjunction (p and q)*
Sequential	*association by order (e.g., 1 precedes 2)*
Temporal	*association by time order*
Communication	*exchange of information*
Procedural	*related by a specified 'way of doing'*
Functional	*bound by a 'way of doing' for purpose*

Coincidental association is the weakest type of cohesion. Logical association is the first meaningful type. Functional binding is the strongest form of cohesion in Yourdon's ranking.

4.3.1 FRAMEWORK FOR STRUCTURED DESIGN

Structured design can be implemented through a framework of model specifications, transformation, and synthesis as depicted in Figure 4.7. Architecture Definition links the technical processes in the upper left corner of Figure 4.1 to Design Definition and proceeds in two steps. Based on the essential definitions, interrelations between the two sets of elements that form the system (elements of the system set and elements of the environment) need to be specified. In the structured analysis process, the

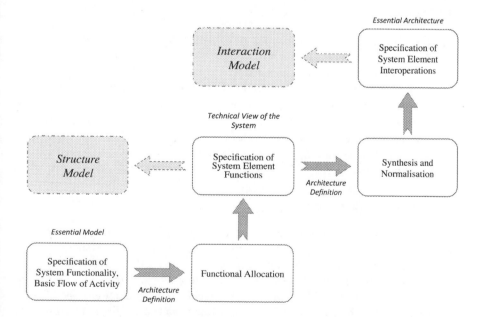

FIGURE 4.7   Framework for Structured Design

behaviours of the system were defined using relations between the system and elements of the environment. These relations were modelled with use cases that were elaborated into functionalities that were actionable. The first step of structured design is to modularise the actions and make them executable.

Various properties can be attributed to the system elements, but it is the functions of the elements that implement the intended actions. These are also called operations, which is the term used in the specification of a UML Class. This type of relation between the actions of the system and operations of the system elements is commonly referred to as *functional allocation*, which is depicted in the lower left Architecture Definition transformation in Figure 4.7. The distinction between functionality and functions (element operations) is that functions are executable as input–output relations. Thus, the term *system element functions* in the figure reflects that system functionality can be implemented in the first-level hierarchy of the system decomposition. If this is not possible then the term should be replaced by *system element functionality*, in which case at least one further level of hierarchy will need to be defined.

In a UML model of the system, the functional allocation will result in a class structure that has implied and intended exchanges between the operations: every input to an element operation must be supplied by another element (either from the system set or the environment), and every output should be consumed by another element. Modularisation of the system elements is accomplished by the UML Classes, which can be implemented in either hardware or software. Class definition of the elements is not specific as to the details of the realisation (e.g., the traditional concept of 'build or buy' at the bottom of the Vee in Figure 4.1). It is not necessary to use UML Classes to specify the elements but as seen in Chapter 3, the graphical notation provides a convenient intuitive method to represent the underlying mathematical classes and logic of a conceptual model of the system.

Before proceeding to the Essential Architecture in the figure, all the previous model specifications must be normalised, i.e., there must be an equivalence of the semantics, syntax, and relations across the concepts and elements of the models that have been specified. This is similar to the traceability in a system and Requirements Definition process. The resulting structure of the models will then reflect not just how the system elements operate and are interrelated but also how they interoperate to achieve the intended purpose of the system. Figure 4.9 in Section 4.4 of this chapter provides an example of an interaction model for the ADAS ECG.

4.3.2 Models for Structured Design

A structured design process can be implemented by model specifications, transformations, and synthesis of the model elements of the system into an interrelational structure for the essential architecture. UML can facilitate the specification of two models that comprise the interrelational structure: the (Class) structure model and the (Object) interaction model. Activity diagrams specified in the essential model of the system are transformed into Class diagrams through a process of functional allocation that associates system actions with system element operations. The interaction model must account for all the required exchanges based on the input–output relations of the operations of the elements. The model is represented in UML (and SysML) by a Sequence diagram that synthesises the system elements (represented by UML Objects), sequencing of operations owned by elements, the interactions (e.g., exchanges) between elements, and the control structure that governs the flow. In terms of UML model elements, the model transformations are realised by the following associations: Class $\rightarrow$ Objects $\rightarrow$ Lifelines; Operations $\rightarrow$ Function Calls. The resulting interrelational structure for the *Essential Architecture* can be refined to specify an *Implementation Model* that enables Design Definition and system implementation.

4.4   REALISING THE SOLUTION CONCEPT FOR THE ADAS ECG

The frameworks for relational specification of models and their transformation, structured analysis, and structured design will now be employed to illustrate how to specify a system that realises the concept for an ECG investigated in Chapter 2. The specification and transformation of first-order models in the relational framework can be used for precise interpretation of the science-based solution into a model-based specification of an information-intensive systems solution. The Use Case diagram in Figure 4.4 serves as a model-based specification of functional requirements. The quantification of these requirements is supported by the response surface analysis organised around Figure 2.1 in Chapter 2. Two objective variables (CO_2 and CO) were analysed in relation to two system characteristics (engine states). A control strategy for how to control the states (engine speed and torque) was developed. In the system solution this becomes a control algorithm that mediates demands from the ADAS to the engine. How this relation is mediated by the ECG is made clear through the intended functionality expressed in the use cases. The development of the algorithm is the concern of a domain engineer. The system structure and requisite properties within which it is implemented is the concern of the architect. This includes the interface requirements for the ECG.

The inputs to and outputs from the ECG are also in the remit of the architect. In addition to functionality and its quantification (based on the use cases, regulatory requirements, and analysis of the constrained engine operating space), the architect must specify the flow of events between the ECG and the ADAS and the engine. For the purpose of this chapter, the synthesis of the essential model of the system represented by an Activity diagram into a Sequence diagram will be sufficient for illustrating the ECG specification as an information-intensive system.

The Activity diagram in Figure 4.8 depicts a design-level model for the system actions of the Engine, ECG, and ADAS that must be performed and the flow of events in their performance. Note that flow does not necessarily imply interaction between system elements. Analyses and design decisions play into the transformation of use cases into a functional flow (flow of actions). The details and demonstration of how this can be done are a subject of Part II tutorials. For this illustrative example, it is sufficient to note the similarities and differences with Figure 4.4. First, the overarching functionality is not explicitly represented in the functional

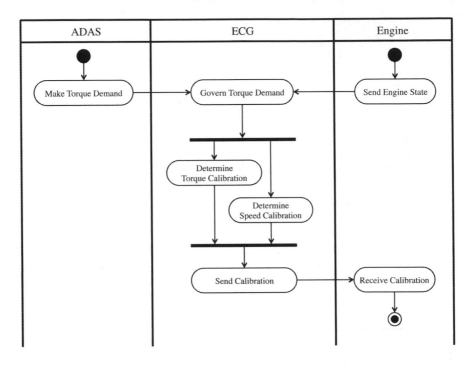

FIGURE 4.8   Essential system model of ADAS ECG

flow. Rather, it is the intended outcome of the flow. Next, the functionality for the ADAS to make a torque demand to the ECG has not changed but the functionality of the ECG to govern torque demand and engine speed has been altered by design decisions. Specifically, the governance of torque demand has been resolved into two calibration actions. These are indicated by the actions between the horizontal bars in the diagram that are symbols for the forking and joining of actions. The actions between these two bars represent an algorithm to be developed by a domain engineer, the output of which is a pair of calibrations that are sent to the engine. Finally, the action to send the engine state to the ECG is not a functionality of the ECG. Rather, it is the outcome of an interface definition. The engine state is needed in order to perform the internal functions of determining calibrations.

The Sequence diagram in Figure 4.9 reflects the structure and system information in the model depicted by Figure 4.8 and adds further detail by specifying the objects (system elements) that will implement the system, along with their operations (functions) and data exchanges. The classes at the top of the 'swim lanes' in the Activity diagram are implemented by one or more of their objects at the top of the so-called lifelines of the Sequence diagram. The flow of actions that snakes horizontally across the diagram in Figure 4.8 and migrates downwards is transformed into a downward vertical sequence of operations along the lifelines in the Sequence diagram. Exchanges between the objects flow horizontally as the events in the environment and the processes of the system unfold.

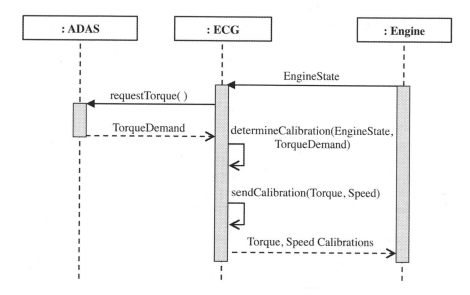

FIGURE 4.9 Interaction model for the essential system architecture of ADAS ECG

In the diagram for the ECG in Figure 4.9, two functions and one function call are specified for the implementation of the actions in Figure 4.8. The algorithm for governing emissions is named "determineCalibration ( )". It is a function of two variables, "Engine State" and "Torque Demand". The other function is the reporting of the calibrations in a format that is suitable for the engine. In the architecture represented in the diagram, the function call from the ECG to the ADAS is a request for the torque demand that the ADAS will be making to the engine. This method is modular in the sense of Structured Design, in that it does not interfere with the routine operation of the ADAS. The dashed arrow is the message with the response from the ADAS to the ECG. This exchange does not need to be implemented directly between the ECG and ADAS. It could also be implemented through a controller area network (CAN bus). In the Sequence diagram, the interactions from the Use Case diagrams are further defined and resolved by interoperations that can identify interfaces between the elements of the system (which are represented as UML Objects). In practice, the three diagrams are specified iteratively and recursively. Because of the interrelations of the underlying models, it is generally not easy to correctly specify all of these simultaneously in the first attempt.

The interrelational structure of the essential architecture, the system models and the engineering models in Chapter 2 provide the minimal level of information that is needed to specify the ECG system for design and implementation. The architect and systems engineer support these next steps in the engineering process but do not design or implement the system elements. Modern Architecture Definition must also support interface specifications for interoperability as part of system integration. The Sequence diagram in Figure 4.9 does this. It also supports specification of a software architecture.

4.5 SUMMARY AND SYNTHESIS

Relations between form and functionality modelled using the laws of science enable the practice of engineering to deliver reliable products and services with market value. The influence of information technology on engineering has become increasingly significant over the first two decades of the twenty-first century in the commercial sector. This has also been fuelled by the development of large-scale information systems in aerospace and defence such as theatre level mission planning systems in which software, hardware, databases, system operators, and increasingly autonomous vehicles are integrated. The importance of system architecture has therefore become more central to modern engineering, as has the need for more formal approaches to system analysis, design, assurance, and specification. Over the same two-decade time period, advanced systems engineering approaches have been expanding from their origin in aerospace and defence to fulfil a need in the commercial sector to deliver increasingly complex system solutions.

If engineering in general and systems engineering in particular are to be founded upon science, then there needs to be a clear understanding as to what is meant by *science*. The scientific method consists of characterisation of observables (by definition and measurement) and statement of hypotheses that are testable and refutable by comparing predictions from models with experimentation (which must be repeatable). The System Requirements and Architecture Definition processes in Figure 4.1 are similar to the characterisation and modelling processes of science. The Verification and Validation processes are similar to scientific experimentation. It should be noted that measurement always involves interaction with the physical object of interest. A scientific arrangement or method of experimentation may be defined as the gathering of individual objects into a synthetic whole. This concept is very close to that of a system. The object of interest in scientific experimentation is a system or element that is isolated from the environment of the experimentation system based on conditions that are considered to be relevant to the outcome. The use case in Figure 4.10 depicts this arrangement, which has a striking similarity with Use Case diagrams in systems engineering. The only substantive difference is the *description* of objects that already exist in the physical world versus the *intent* of objects that are designed and developed in engineering.

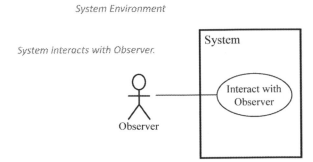

FIGURE 4.10 Use Case diagram for scientific experimentation and discovery

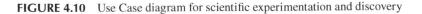

The similarity between science and engineering is an example of logical homology; that is, a congruence that provides guidance for understanding the underlying causes of similarity. Although scientific method is concerned with explanation of observation, homologies have provided useful models in science (Bertalanffy 1969). The Use Case diagram in UML is a logical homology, especially as used in this book in terms of verb-noun predicates. An example in biology is the similarity of cells due to shared ancestry between a pair of structures or genes. In physics, another example is the mathematical similarity of models for heat and electricity as flows of a substance despite their substantial physical differences. Therefore, engineering has more in common with science than just using its laws and models. Logical homology is also used in this book in the similarity of the models in the tutorials in Part II and those of the case studies in Part IV. Over a half century ago, the noted biologist and theorist Ludwig von Bertalanffy considered that any general theory of systems should be able to distinguish between analogies (patterns or other superficial similarities), which are regarded as useless by science, and homologies which, on the other hand, can be useful.

The need for more formal approaches to the architecting and engineering of modern systems demands that concepts and terminology be defined with greater precision, yet remain practical and intuitive for comprehensibility. Formal approaches are a subject of the first-order model theory that underlies the mathematical expression of models in modern science. Modelling is interpretation of concepts into structures that represent the system. This applies to both science and engineering. The definition and lack thereof for key terms in the current relevant international standards generally does not provide the precision required by a rigorous approach to architecture and systems engineering. To fill this gap, essential definitions based on recent summative research (Dickerson et al. 2021) have been offered in Chapter 3 for the terms structure, architecture, system, and engineering. These definitions are used in the essential technical processes and structured methods put forth in this book. They are intended to complement the standards, and not to be a replacement.

The precision and expressive power of the essential definitions are seen in the unified model depicted in Figure 4.11. The synthesis of concepts in the figure extends the logical model of system based on standards in Figure 3.1 of Chapter 3 into the unified model. The six natural language predicates in Figure 3.1 have been synthesised with the predicates of the essential definitions to form a holistic logical model of the eleven natural language predicates depicted in the unified model.

Figure 4.11 can be used, for example, to explain the relationship and distinction between *system* and *architecture*. The current standards offer no such normalisation of these key terms; and as discussed in Chapter 3, this leads to confusion about the relation between architecture and models in the standards, amongst other things. In the unified model, the definition of system is seen to be a *first-order logical statement* that relates the elements of a defined set of system elements with those of an environment. It can be modelled using only the entities and their first-order relationships in Figure 4.11 (represented by nouns that name abstract classes and the relationship lines between them). On the other hand, the definition of system architecture is seen to be a *second-order logical statement* that relates constraint properties with

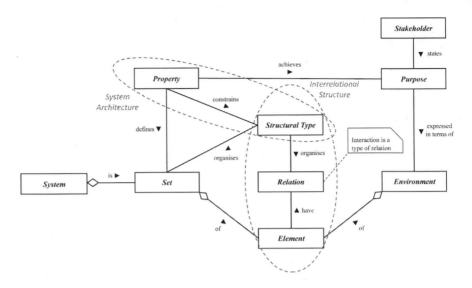

FIGURE 4.11   Synthesis of concepts for system and architecture into a unified model

structural type; it is a relation between properties and relations rather than a relation between entities. (Recall that structure is a relation between the objects of a class.)

This insight has broad implications. One is that the Essential Model of a system produced by the Structured Analysis process and methods in Section 4.2 is indeed a first-order model in the sense of predicate logic and Tarski model theory. Another is that an Essential Architecture of a system produced by the Structured Design process and methods in Section 4.3 specifies an interrelational structure that has been constrained by the Essential Model and various properties such as specified functionalities, performance, constraints from non-functional requirements, and other architectural properties. In general, there is no first-order model of Architecture. This was foreshadowed by the abstract (predicate) representation of the relational framework in Figure 4.5 that underlies the methods of Structured Design used in this book. Recall that in the formal framework, predicates and (logical) sentences are transformed into a predicate structure, i.e., a structure of relations between relations. The reason a Sequence diagram such as in Figure 4.9 can be represented in UML is that UML is not a formal language, despite its roots in class theory and logic. The complexity of the interrelationships in the Essential Architecture will be discussed in the second tutorial (Chapter 6). It should be further noted that the distinction of *system* as a first-order concept in the diagram in Figure 4.11 and *architecture* as a second-order concept also resolves confusions that are a subject of debate in the international community even at the time of the writing of this book.

Chapter 4 has shown how systems, architecting, modelling, and engineering can be bound into a straight-forward process implemented by rigorous methods for architecture and system definition of solutions to complex problems. This fulfils a need for methods that implement an easily accessible yet rigorous end-to-end process to exploit the advances that have been made in tools and languages. This goes beyond traditional system engineering methods. Structured Analysis and Design, for example, have been explained in the context of current standards and modelling languages using a widely accepted core of object-oriented graphical models: Use Case diagrams, Activity diagrams, Class diagrams, and Sequence diagrams. These models express a concept of integration (based on compliance to architecture) for modern systems that goes beyond physical interfaces and connecting parts. Integration today must also include a blueprint for system interoperability. These ideas have been illustrated through the elementary example of the ECG that has been presented in bite-sized steps using a thread of discussion that began in Chapter 2. The mathematical language and methods of science and model theory have been expressed in a straightforward intuitive engineering process using state of the art object-oriented graphical models. The conceptual integrity of the system can therefore be maintained throughout the development process by a comprehensive architecture that enables the design and development of hardware and software in a holistic concordant way.

Part I Bibliography

Bell, John Lane, and Alan B. Slomson. 1969. *Models and Ultraproducts*. Amsterdam: North Holland.

Brooks Jr Frederick P. 1995. *The Mythical Man-month*, Boston: Addison-Wesley Longman Publishing.

Cameron, Bruce, and Daniel M. Adsit. 2020. "Model-based systems engineering uptake in engineering practice." *IEEE Transactions on Engineering Management* 67, no. 1: 152–162.

Chen, Peter Pin-Shan. 1976. "The entity-relationship model – toward a unified view of data." *ACM Transactions on Database Systems (TODS)* 1, no. 1: 9–36.

Cloutier, Robert and Mary Bone. 2015. "MBSE Survey." Presentation given in INCOSE IW Los Angeles, CA. January 2015. http://www.omgwiki.org/MBSE/lib/exe/fetch.php?media=mbse:incose_mbse_ survey_results_initial_report_2015_01_24.pdf

Dickerson, Charles E. 2008. "Towards a logical and scientific foundation for system concepts, principles, and terminology." In *2008 IEEE International Conference on System of Systems Engineering*, (June 2008): 1–6.

———. 2013. "A relational oriented approach to system of systems assessment of alternatives for data link interoperability." *IEEE Systems Journal* 7, no. 4: 549–560.

Dickerson, Charles E, and Dimitri N. Mavris. 2010. *Architecture and Principles of Systems Engineering*. New York: CRC Press.

———. 2013. "A brief history of models and model based systems engineering and the case for relational orientation." *IEEE Systems Journal* 7, no. 4: 581–592.

Dickerson, Charles E, and Siyuan Ji. 2018. "Analysis of the vehicle as a complex system, EPSRC." *Impact* 2018, no. 1: 42–44.

Dickerson, Charles E, Siyuan Ji and David Battersby. 2018. Calibration system and method. U.K. Patent GB2555617, filled November 4, 2016, and issued May 09, 2018

Dickerson, Charles E, Stephen M. Soules, Mark R. Sabins, and Philipp H. Charles. 2003. *Using architectures for research, development, and acquisition*. ADA427961. Prepared by Office of the Chief Engineer of the Navy, Assistant Secretary of the Navy. Defense Technical Information Center (www.dtic.mil).

Dickerson, Charles E, Michael Wilkinson, Eugenie Hunsicker, Siyuan Ji, Mole Li, Yves Bernard, Graham Bleakley, and Peter Denno. 2021. "Architecture definition in complex system design using model theory." *IEEE Systems Journal* 15, no. 2: 1847–1860. doi: 10.1109/JSYST.2020.2975073.

Dori, Dov. 2016. *Model-Based Systems Engineering with OPM and SysML*, New York: Springer.

Estefan, Jeff A. 2007. "Survey of model-based systems engineering (MBSE) methodologies." *INCOSE MBSE Focus Group* 25, no. 8: 1–12.

Finch, James K. 1951. *Engineering and Western Civilization*. New York: McGraw Hill.

Hatley, Derek, Peter Hruschka, and Imtiaz Pirbhai. 2000. *Process for System Architecture and Requirements Engineering*. New York: Dorset House Publishing.

Hawking, Stephen. 1988. *A Brief History of Time*. New York: Bantam Books.

Hilliard, Richard F., Timothy B. Rice, and Stephen C. Schwarm. 1996. "The architectural metaphor as a foundation for systems engineering." In *INCOSE International Symposium* 6, no. 1: 559–564.

Hodges, Wilfrid. 2020. "Model theory," In *Stanford Encyclopedia of Philosophy*, Online ed., edited by Edward N. Zalta. Stanford: Metaphysics Research Lab, Stanford University. https://plato.stanford.edu/entries/model-theory/

IEEE (Institute of Electrical and Electronics Engineers). 2000. *IEEE recommended practice for architectural description for software intensive systems*. IEEE Std 1471-2000.

INCOSE (International Council on Systems Engineering). 2015. *Systems Engineering Handbook*. 4th ed. New Jersey: John Wiley & Sons.

ISO (International Organization for Standardization). 2001. *System and software engineering – System life cycle process*. ISO/IEC/IEEE 15288:2002. Standard superseded by ISO/IEC/IEEE 15288:2015.

———. 2002. *System and software engineering – System life cycle process*. ISO/IEC/IEEE 15288:2002. Standard superseded by ISO/IEC/IEEE 15288:2015.

———. 2011. *Systems and software engineering – Architecture description*. ISO/IEC/IEEE 42010:2011.

———. 2015. *System and software engineering – System life cycle process*. ISO/IEC/IEEE 15288:2015.

———. 2018. *Information technology – Common Logic (CL) – A framework for a family of logic-based languages*. ISO/IEC 24707:2018.

Klir, George J. 1991. *Facets of Systems Science*. New York: Plenum Press.

Lin, Tzu-Chi, Siyuan Ji, Charles E. Dickerson, and David Battersby. 2018. "Coordinated control architecture for motion management in ADAS systems." *IEEE/CAA Journal of Automatica Sinica* 5, no. 2: 432–444.

Lin, Yi. 1999. *General Systems Theory: A Mathematical Approach*. New York: Kluwer Academic/Plenum Publishers.

Lin, Yi, and Yong-Hao Ma. 1987. "Remarks on analogy between systems." *International Journal of General System* 13, no. 2: 135–141.

OMG (Object Management Group). 2003. *MDA Guide Version 1.0.1*. Edited by Joaquin Miller and Jishnu Mukerji. OMG document omg/2003-06-01. OMG

———. 2014. *MDA Guide Version 2.0*. OMG document omrsc/2014-06-01. OMG

———. 2017a. *OMG Systems Modeling Language (SysML®)*, Version 1.5.

———. 2017b. *Unified Modeling Language (UML®)*, Version 2.5.1.

———. 2019. *UML Profile for MARTE (MARTE)*, Version 1.2.

Pyster, Art, Rick Adcock, Mark Ardis, Rob Cloutier, Devanandham Henry, Linda Laird, Michael Pennotti, Kevin Sullivan, and Jon Wade. 2015. "Exploring the relationship between systems engineering and software engineering." *Procedia Computer Science* 44, no. 2015: 708–717.

Rosen, Robert. 1993. "On models and modeling." *Applied Mathematics and Computation* 56, no. 2–3: 359–372.

Sowa, John F. 1983. *Conceptual Structures: Information Processing in Mind and Machine*. Reading: Addison-Wesley.

———. 2000. *Knowledge Representation: Logical, Philosophical, and Computational Foundations*. Brooks: Cole Pacific Grove.

Tarski, Alfred. 1954. "Contributions to the theory of models. I & II." *Indagationes Mathematicae (Proceedings)* 16, no. 1954: 572–588.

———. 1955. "Contributions to the theory of models. III." *Indagationes Mathematicae (Proceedings)* 17, no. 1955: 56–64.

Von Bertalanffy, Ludwig. 1967. "General theory of systems: Application to psychology." *Social Science Information* 6, no. 6: 125–136.

———. 1969. *General Systems Theory*. New York: George Braziller Inc.

Wasson, Charles S. 2005. *System Analysis, Design, and Development: Concepts, Principles, and Practices*. Hoboken: John Wiley & Sons.

Wymore, A. Wayne. 1993. *Model-based Systems Engineering*. Boca Raton: CRC Press.

Yourdon, Edward. 1989. *Modern Structured Analysis*. Upper Saddle River: Yourdon Press.

Part II

Tutorial Case Study: End-to-End Demonstration

OVERVIEW OF TUTORIAL CASE STUDIES

The purpose of the three tutorial chapters in this part is to demonstrate an elementary end-to-end example of how the essential technical processes and methods in Part I can be applied to the model-based development and specification of a system solution to meet a customer need for improving mission capabilities. They significantly extend the practices used in the simple example of the emissions control governor in Part I. Each chapter in Part II demonstrates one aspect of the system development and specification process. A summary of the models specified and the transformation between them is provided at the end of each chapter.

A familiar example requiring a minimum of technical detail has been chosen for this purpose: an air traffic control system at a local airport. This is intended to make the tutorial accessible and interesting to a broad audience. The solution developed will be an information-intensive system with both a hardware component (a radar system) and a software component (an air traffic management system). Formative exercises are provided throughout each chapter to engage the reader in general and students in particular in the thought process of developing the system solution. The progression of the three chapters in the tutorial leads the student in bite-sized steps to the system solution and its specification in terms of a comprehensive architecture and set of essential models of the system.

SYSTEM CONCEPT AND CONTEXT

Air traffic control is a service that manages the relation between an airport and participating aircraft. Basically, each aircraft seeks to enter the airspace, make a safe approach to the runway, and land safely on time. The airport can be regarded as a business that seeks to maintain a sufficient flow of aircraft to profitably sustain its operations. The one in the case study can be thought of as a local airport that is seeking to upgrade the equipment of an aging air traffic control service. The system developer on the other hand sees this as an opportunity not just to sell a 'technology refresh' to the customer but rather to sell new mission capabilities that support an increased flow of traffic through the airport and deal with a safety issue that has been neglected in the past.

THE ROLE OF THE ARCHITECT

Developers for modern systems frequently employ a system architect to assist the marketing team in negotiations with the customer and ensure that agreements between the customer and developer can actually be delivered by the engineering team. As noted in Part I and put forth by Brooks, *The architect should be responsible for the conceptual integrity of all aspects of the product perceivable by the user. Conceptual integrity is the most important consideration in system design.* Although the System Architect is customer and user-focused, the outputs of system architecting and engineering must include specifications to the subsystem engineers that are in sufficient detail for them to develop the subsystems and their interfaces. These specifications must derive from a comprehensive architecture and set of essential models of the system that can be realised as products and services by the engineering team. They must be traceable back to the customer needs and mission requirements. A model-based specification for the design and system implementation of an air traffic control system proposed by the developer will conclude the tutorial case studies in this part of the book.

5 Tutorial I
Air Traffic Control System Essential Model

KEY CONCEPTS

System functionality
Definition and decomposition
Basic flow of actions
System behaviour

The purpose of this tutorial is to illustrate the *practice* of essential activities in a System Definition technical process. It shares a parallel structure with and gives insight into the first part of the practical case study (Chapter 11, Section 11.1) but is not intended to solve it. The essential activities and methods that implement them provided in Chapter 4 are aligned to key systems standards that were reviewed in Chapter 3. The practical activities of systems definition presented in this tutorial are:

Transformation of concepts and user needs into system requirements
An interaction-based approach to modelling system functionality
Transformation to system behaviour model
Specification of the essential model for the system

The models produced in this tutorial will be expressed in terms of the Unified Modeling Language (UML) diagrams that were used in Chapter 4 and will be reviewed in detail in Chapter 8 on modelling languages. Key definitions of terminology for this tutorial are summarised in Chapter 3 and Annex A-3.

5.1 TRANSFORMATION OF CONCEPTS INTO REQUIREMENTS

The specification of requirements is one of the key outputs of systems engineering. This becomes a contractual basis for both developing and delivering a product or service, i.e., a system that has market value. The technical processes of developing a requirements specification belong to an overarching process that is referred to as *requirements definition* or *requirements engineering*. System requirements are generally classified as being either *functional* or *non-functional*. This is an example of systems thinking that separates the *what* from the *how*. Functionality is the purpose a system is intended to fulfil. So, functional requirements are concerned with what a system must do to achieve its intended purpose. The so-called non-functional requirements, on the other hand, are concerned with how well the system will provide functionality and how it will be implemented. Requirements definition in a top-down systems engineering process begins with eliciting requirements from the stakeholders of the system. This is facilitated by a narrative about the system that should be a common focal point between the acquiring stakeholder and the system developer. The scope of concerns in this tutorial start at the point where requirements elicitation has been completed to a level of detail that discussions can begin with the system developer.

Requirements definition is part of an iterative process of refinement and agreement between the developer and the acquiring stakeholders. Ambiguities and so-called requirements creep can put the project at risk. The architect must be an active partner in the requirements definition process. The behavioural focus in the tutorial is aimed at specifying a stable layer in the system architecture to isolate and manage these types of risks. Model definition in terms of classes (i.e., types of objects) rather than specific objects is also a method for this type of risk management.

The system of interest for all three tutorial chapters in this part of the book will be an air traffic control system for commercial aviation that operates within a system of airports. The following narrative is provided and has been 'marked up' to identify key words, concepts, and constraints. It can be regarded as part of a briefing package that will be an on-going refinement of concepts between the acquiring stakeholder and the system developer.

> **Air traffic control (ATC)** is an essential **service provided** by Class A and B **airports**. The **service** will be implemented by a controller at the airport (who is referred to as the ATCC) and an Air Traffic Control System (ATCS) comprised of an **Air Traffic Radar (ATR)** and a **system for flight management**. The airport is part of an air traffic system of systems (ATSoS) that is comprised of participating aircraft, airports, a regional flight plan management system, and infrastructure to include **communication systems** linking aircraft to each other and the ground facilities. The ATCS must support **the peak tempo of the airport**. Occasionally there are unintended aircraft in the ATC airspace. The ATCS is governed and influenced by the civil aviation agency, standards and regulatory organisations, and local government.
>
> The ATCS needs all-weather tracking capabilities to accomplish its purpose. **Radar** is the favoured technology for all-weather tracking.

The Use Case diagram in the figure provides an initial depiction of the system boundary of the ATC service. In this way, the service is conceptualised as a system. The actors belong to the system environment and not to the system itself. The diagram expresses how the system is used in terms of its interactions with the environment. At this level of description, the system is a black box. The two high-level use cases provide a starting point for specifying what is generally referred to as the functional architecture of the system. The functionality will be refined through repeated application of functional decomposition of the use cases until a complete set of system functions is defined. Note that the narrative at this point is silent on how the functionalities of the ATC are performed. The system developer seeks further understanding from the stakeholders on this point, especially as this might be an opportunity to sell more than a radar to the customer.

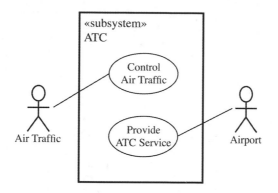

FIGURE 5.1   ATC Use Case diagram: An Airport Service

Exercise 5.1

1. Using the Use Case diagram in Figure 5.1 as a reference, draw a high-level Use Case diagram for the ATCS.
2. Explain your rationale for the use cases.

5.2   MODELLING SYSTEM FUNCTIONALITY

A more detailed narrative is needed that expresses the operational activities of the ATCS to define the system functionality at a level of description that is useful both to the marketing team and the engineers who must develop and deliver the system, i.e., 'realise the concepts'. The following narrative of the operational concept is provided and has been 'marked up' to identify key words and activities.

> The purpose of the ATC service is to manage the flow of aircraft through the airport airspace and to prevent accidents (especially air-to-air collisions). The **operational activities** of the ATCS include:
>
> * **Track the aircraft** entering and departing the controlled airspace.
> * **Assess the flight paths** against: (i) **the flight plans**, (ii) **the tempo of the airport**, and (iii) **the risk of air-to-air collisions**.
> * Control the aircraft to **manage their relative positions and flow** through the airspace.
>
> The **ATR must be suitably positioned**, e.g., at the airport to provide tracking of all aircraft entering and departing the controlled airspace (which is also referred to as the ATC airspace).

This more detailed narrative will be used to elaborate the high-level functionality of the ATCS based on the ATC functionality expressed in Figure 5.1. When interpreting the information in the narrative into Use Case diagrams, it is important to preserve the terms, their semantics, and their relations as expressed in the natural language statements in the narrative.

Figure 5.2 is an example of such an elaboration. The concept of 'unintended aircraft' in the part of the narrative in the previous section has been expressed

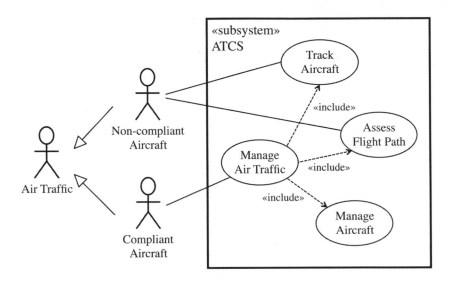

FIGURE 5.2 ATCS Use Case diagram: Manage Air Traffic (functional definition and decomposition)

using two types of aircraft in the air traffic: compliant and non-compliant. The compliant aircraft are those aircraft intended to be in the ATC airspace and therefore are assumed to comply with the flight regulations to be there. For example, these aircraft are equipped with a transponder that is used for identification.

The diagram in the figure refines the ATC "Control Air Traffic" use case by the method of definition and decomposition; in this case, applied to the functionality of the system (not its component elements). Non-compliant aircraft are not managed in this construct. The terms in the narrative have been preserved but distilled into verb-noun phrases appropriate for each use case. Note however that at this point, the relationship between the ATC and ATCS has not yet been expressed in terms of the use cases. In the diagram in Figure 5.2, the system developer is seeking to bring some of the functionality of the ATC service into their proposed ATCS solution.

From this point forward in the ATCS, it will be helpful to narrow the scope of discussion to maintain a simple train of thought. The purpose of the tutorial is *to illustrate the practice* of essential activities in System Definition and not to go into unnecessary depth. To this end, the scope of the functional requirements analysis will be limited to the following ATC operational activities: track, assess, and manage. Further simplifying assumptions will be made. The ATCS will be concerned only with approaching aircraft. Also, if the ATR is to be part of the airport ground facilities, then it will only be required to stare along the approach to the runway, i.e., it will not be required to search the airspace.

Exercise 5.2

1. Draw a separate Use Case diagram at the 1st level of decomposition for the ATR based on the "Track Aircraft" use case. This includes detection and identification of aircraft.

2. Do the use cases in Figure 5.2 imply that the ATR must:
    a.　Detect non-compliant aircraft?
    b.　Identify non-compliant aircraft?
3. The narrative refers to a second component to the ATCS that is a system for flight management to support the ATC Controller. Based on the Use Case diagram in Figure 5.2, partition the uses cases of those to be performed by the flight management system from those to be performed by the ATR. The second component of the ATCS will be given the name Air Traffic Management System (ATMS) and proposed by the system developer as a new development item.
4. Use the operational activities in the narrative to further define and decompose the use cases of the ATCS in Figure 5.2 into a second level of functionality for the proposed ATMS.

5.3　SYSTEM BEHAVIOUR MODEL

As previously noted, functionality is the purpose a system is intended to fulfil; and functional requirements are concerned with what a system must do in order to achieve its intended purpose. The modelling of functionality and functional requirements with use cases is part of specifying what a system does. Behaviour in science is regarded as a change in state such as a change in the value of an attribute. Engineered systems also have processes. At the most fundamental level, the state of a process is its status, i.e., whether the process is active or idle. The UML diagram that is used for modelling this type of state change is the Activity diagram. It arranges the process states of a system into a flow of being active. These can be semantically linked to the use cases to give a holistic view of the system functionality and behaviour.

The elaborated Use Case diagram in Figure 5.2 provides such a link because it was based on the operational activities of the ATCS. This provides a starting point for specifying an Activity diagram for the ATCS at the system level (i.e., without reference to the two given components that it is comprised of). A popular method is to organise the information in a Use Case diagram into a table that is referred to as a Use Case description that expresses the use case in the context of a scenario. Further information is integrated, such as pre- and post-conditions. A basic flow of actions associated with the use case is also listed. (Extension points and alternate flow will be discussed in the third tutorial [Chapter 7].) Table 5.1 provides a Use Case description for the "Manage Air Traffic" use case. The scenario includes specialisation of "Air Traffic" to "Compliant Aircraft". New relations have been added: for example, a precedence relation between "Assess Flight Path" and "Manage Aircraft". In the basic flow of actions, the flight path of a compliant aircraft must be assessed before it can be managed. This is also evident in the operational activities that were detailed in the narrative. It is also evident in the narrative that the aircraft controlled (by the ATC) and management of the aircraft refer to air traffic flow. Additional new information could be based on either a design decision about the ATCS or further description from domain knowledge.

Based on the scenario in the table and following the construct of the system boundary in the Use Case diagram in Figure 5.1 (and Exercise 5.1), the ATCS and its environment are partitioned into two flows of actions, which are referred to as swim

TABLE 5.1

ATCS Use Case Description: Manage Air Traffic (Scenario 1: Compliant Aircraft)

Use Case Name	**Manage Air Traffic (Scenario 1)**
Description	Manage approaching Compliant Aircraft
Actors	Compliant Aircraft
Pre-conditions	Aircraft request permission to enter ATC airspace
Post-conditions	Aircraft cleared for landing
Actions for Basic Flow	1. Track Aircraft
	2. Assess Flight Path
	3. Manage Aircraft
	4. Track Aircraft
Extension Points	None
Alternate Flow	None

lanes in an Activity diagram. The terms from the Use Case description along with their semantics and relations are then embedded into the diagram and the swim lanes. Further domain knowledge is integrated to develop a more complete description of the system-level flow of actions. For example, flight regulations require that aircraft request permission from the ATC before entering controlled airspace. Figure 5.3 depicts an Activity

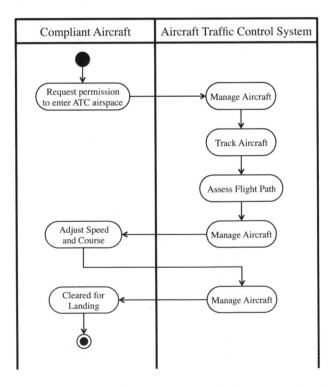

FIGURE 5.3   ATCS Activity diagram: manage air traffic (Scenario 1: Compliant Aircraft)

diagram that is the outcome of embedding information in a Use Case description into an Activity diagram. This is a high-level representation of the system behaviour of the ATCS as a whole but is mostly concerned with air traffic management.

Definition and decomposition can be applied to the ATCS in terms of its two components identified in the narrative (i.e., an ATR and ATMS as specified in Exercise 5.2.3). A Use Case description for the ATR can also be specified based on the Use Case diagram for the ATR (from Exercise 5.2.1). The basic flow for the ATR can then be integrated into the Activity diagram for the ATCS (Figure 5.3) and further structured as depicted in Figure 5.4. The actions of the ATR and the ATMS are separated by the swim lanes but the specification of precedence relations must be across the swim lanes and not just down them. These details will be critical to the system design engineers. However, at this stage of system definition, there is still ambiguity as to what the ATMS is and how it relates to an air traffic controller in the ATC.

The diagram in Figure 5.4 is a high-level representation of ATCS behaviour as a system of two components for the compliant aircraft scenario. It is important to keep in mind that an Activity diagram only shows order and separation of actions. The diagram as depicted in the figure is correct but does not imply that the ATMS actually 'controls' the compliant aircraft. This is done by the ATC, as indicated in Figure 5.1

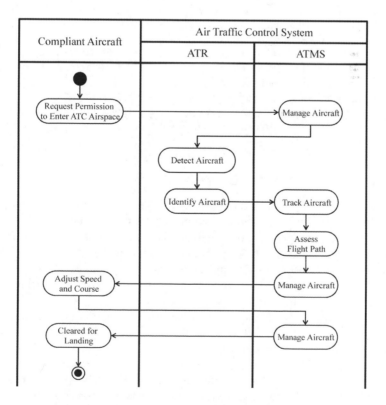

FIGURE 5.4   ATCS Activity diagram: Model of System Behaviour (Scenario 1: Compliant Aircraft)

through the "Control Air Traffic" use case. Clarification could be improved by adding a swim lane for the ATC to the diagram, e.g., on the right and side. The arrow in Figure 5.4 from "Request Permission to Enter ATC Airspace" to "Manage Aircraft" would be redirected to an action in the right-most swim lane (for the ATC) such as "Grant Permission". A new arrow would then need to be drawn back to "Manage Aircraft" before returning to the "Adjust Speed and Course" action for the compliant aircraft.

The key functionalities of the proposed system solution are detection (and track) of arriving aircraft and management of the aircraft under peak tempo conditions. In the activity model in Figure 5.4, each of these functionalities has been allocated to one of the proposed ATCS subsystems. These can be quantified by metrics such as the detection range of the radar (e.g., 30 nmi) and flow rate of the aircraft (e.g., one aircraft landing every 2 minutes).

Exercise 5.3

1. Following the paradigm of use case → functional decomposition → Use Case description → Activity diagram, what is needed to complete the transition from Figures 5.3 to 5.4?
2. Based on the first level decomposition in the Use Case diagram for the ATR (from Exercise 5.2.1), specify a Use Case description for the ATR in the compliant aircraft scenario. This complements the Activity diagrams in Figures 5.3 and 5.4.
3. Based on the elaborated Use Case diagram for the ATCS in Figure 5.2 and using Table 5.1 as a reference, specify a Use Case description for the non-compliant aircraft scenario.
4. Based on the Use Case description for the ATCS in the non-compliant aircraft scenario in the previous exercise and using Figure 5.3 as a reference, specify an Activity diagram for the ATCS in the non-compliant aircraft scenario.

5.4   AIR TRAFFIC SYSTEM OF SYSTEMS

At this early stage of design and development, there will likely be ambiguities that need to be resolved before the project can move forward into contractual agreement. At the system level of specification, the relation between the ATC service and ATCS needs further clarification. At the system of systems level, the narrative is silent on how the functionalities of the ATC are performed. The system developer should seek further understanding from the stakeholders on this point, especially as this might be an opportunity to sell more than a radar to the customer.

The ATSoS description in the customer narrative provides contextual information that can be useful for addressing some of the ambiguities as well as an opportunity to explore a more extensive development of the ATCS concept in the design proposal. For example, the flight plan management system necessarily operates across all the airports in the system of systems. It might be an enabling system that could inform the ATCS of the larger air traffic context. The question could then be raised regarding how the ATCC uses the ATSoS planning system and whether it too might be exploited better through integration with the ATCS. This might support a viable solution that is more effective and efficient in supporting the ATCC at the local airport. The ATSoS concepts could then become a subject of discussion between the acquiring stakeholder and the system developer.

5.5 SPECIFICATION OF THE ESSENTIAL MODEL

At this stage of an engineering project to deliver an ATCS to an acquiring stakeholder, the architect has developed an initial technical package that is suitable for a review meeting with the stakeholders. The higher-level UML diagrams would be used in an executive-level presentation whilst the more detailed diagrams would be intended for the technical team of the stakeholder. These can also be used to address specific questions of detail that might arise during the executive presentation.

The purpose of the meeting would be to seek agreement on the concepts of and functional requirements for the system. Traceability to the narrative demonstrates to the stakeholders that the systems engineering team is listening to the voice of the customer. Depending on the acquisition process of the acquiring stakeholder, a request for proposal (RFP) or request for information (RFI) might have been issued. These would be used in the same way that the narrative has been in this tutorial chapter. The RFP or RFI might also include such a narrative. Ideally, the architect would also be in a position to argue that the technical package has been produced through a qualified model-based systems engineering process.

The technical package for a preliminary system specification would be comprised of the dozen or so UML diagrams from the figures, tables, and exercises in this chapter. These should be complemented with succinct explanatory text. The executive summary of the package could be presented as a check list that indicates project status and points out key items for the next step in design and development. Here is an example based on the progress of this first tutorial chapter:

- An essential model of the system has been completed that includes:
  - ATCS purpose, boundary, elements, functionality, and behaviour
  - Initial identification of enabling systems, infrastructure in the ATSoS
- System requirements have been specified for high-level functionality
  - Track Aircraft, Assess Flight Paths, Manage Aircraft
  - Functional decomposition resulting in 7 more detailed requirements
- High-level system behaviour has been specified in two key scenarios
- System requirements have been analysed
- Apart from needing to agree upon the various details of the system concept as viewed by the system developer, there is a significant issue regarding what functionality of the ATC service will be allocated to the ATCS and what will be in the sole remit of the ATCC.

Traceability to the narrative has been ensured by the functional definition and decomposition process. The architecture team is ready to seek agreement between stakeholders and engineering *but need agreement on the relation between the ATCS and ATCC; the ATSoS concept, and an objective for the airport peak tempo.*

- Technical details, requirements, and domain knowledge have been expressed in terms of UML graphical models to facilitate precision, communication, and a forthcoming specification of a holistic architecture for software and hardware development. These have been produced by rigorous model-based systems engineering technical process that is aligned to modern standards.

5.6 SUMMARY OF MODEL SPECIFICATIONS AND TRANSFORMATION

The chapter summary is comprised of two parts. The first is a refinement of the system concept provided in the stakeholder narrative that is briefly summarised from the viewpoint of the system developer. The second is a specification of the types of models used in the chapter and their transformation.

5.6.1 REFINEMENT OF THE SYSTEM CONCEPT

The current (as-is) ATC service has been implemented by an ATCC and an ATR. These two elements form the first-level hierarchical decomposition of the service components. Two key actors were also specified in the high-level description of the service: Air Traffic and the Airport that hosts the service. (These can be thought of as clients of the ATC service.) These four elements defined the initial system boundary of the service.

The customer is seeking to improve the service by implementing an Air Traffic Control System (ATCS) comprised of the ATR and a system for flight management. There is also a flight plan management system (FPMS) in the system of systems that the airport belongs to but details of the FPMS have not been provided or investigated at this stage of the proposal process. The customer also made no indication as to whether the ATR needs to be upgraded or if they are interested in replacing it with a new radar. As a component of the ATC service, the ATCS interacts with air traffic and the ATCC. In the tutorial, these four elements (two actors and two system components) defined the initial system boundary of the ATCS. Two types of air traffic were also identified in the initial ATCS concept: compliant aircraft and non-compliant aircraft.

The primary focus of the analysis at this stage of the proposal has been on the system for flight management which traces to the key functionalities defined for the ATC service: to track and manage the aircraft, and to assess their flight paths. Performance of the functionalities need to be quantified by metrics such as detection range of the radar and tempo of the airport, e.g., 30 nmi and one aircraft landing every 2 minutes. All-weather operation was a non-functional requirement. Regulations also need to be investigated further. The system developer is seeking to integrate ATC track management functionalities into the ATCS through a system for flight management and define how it should integrate with the ATR to best support the ATCC.

5.6.2 MODEL TYPES AND TRANSFORMATION

Chapter 4 introduced processes and structured methods for System Definition. Two essential types of models were specified in terms of UML: Use Case and Activity diagrams. These were sufficient to define the system boundary, functionality, and behaviour. Functional decomposition, as well as requirements traceability is accomplished by model transformation. The tutorial in this chapter has implemented the Framework for Structured Analysis that was depicted in Figure 4.6. The three stages of model specification and transformation in the tutorial are summarised generically in Figures 5.5–5.7 then mapped into the behavioural model in Figure 5.8. Specifically, these are:

- Specification of system functionality
- First-level functional decomposition
- Specification of Use Case descriptions
- Mapping of descriptions into basic flow of actions

Figure 5.5 illustrates a transformation of an external view of the system boundary into use cases. Associations between actors in the environment and the system (as a black box) are based on interactions. In the tutorial, the users in the environment of the ATC were the airport and the participating aircraft. The stakeholder needs were expressed at a primitive level by verb-noun phrases that defined the use cases: "Provide ATC Service" and "Control Air Traffic".

Functional decomposition was accomplished by specifying included use cases. The "Manage Air Traffic" use case was decomposed into "Track Aircraft", "Assess

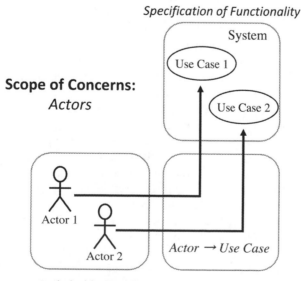

FIGURE 5.5 Specification of system functionality

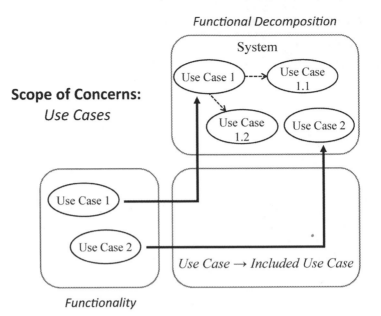

FIGURE 5.6   First-level functional decomposition

Flight Path", and "Manage Aircraft". The control of the aircraft was allocated to the ATCC. Based on the narrative, the two types of aircraft were defined as: compliant aircraft (that were participating in the control process) and non-compliant aircraft (which were identified as unintended aircraft in the airspace by the narrative).

When the functionalities in the use cases were decomposed to a level that was actionable, they were associated with actions that formed the basic flow of actions through the system. These became the processes of the ATCS in a baseline scenario in which the system would operate under intended conditions. They were derived from the operational concept provided by the customer. The key elements of a scenario are the initial state, the intended end state, the course of actions (i.e., behaviour) to reach the end state, and the entities (actors and systems or subsystems) together with their relations. An operational scenario is a coarse description of the intended situation and its dynamics. Alternative scenarios reflect changes to the intended situation. In the tutorial, this was a change in the type of aircraft (i.e., compliant → non-compliant). These changes might require alternative behaviours from the system. (Refer to A-3.2.11 in Annex A-3 for further details about the term scenario.)

The graphical organisation of the information in the use case decomposition was then entered into a structured table that captured scenario details but separated them from behavioural details. The outcome of this process is depicted in Figure 5.7. A line for extension points in the table is provided in preparation for dealing with alternative behaviours, such as alternate flows for dealing with non-compliant aircraft. These will be discussed further in the next two tutorials (Chapters 6 and 7). This type of diagram is referred to as a Use Case description. An example was provided in Table 5.1 for the "Manage Air Traffic" use case in the compliant aircraft scenario.

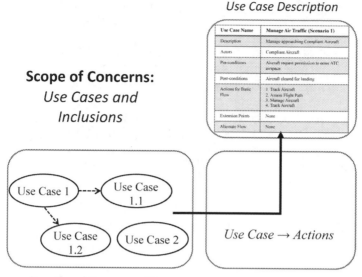

Use Case Description

Use Case Name	Manage Air Traffic (Scenario 1)
Description	Manage approaching Compliant Aircraft
Actors	Compliant Aircraft
Pre-conditions	Aircraft request permission to enter ATC airspace
Post-conditions	Aircraft cleared for landing
Actions for Basic Flow	1. Track Aircraft 2. Assess Flight Path 3. Manage Aircraft 4. Track Aircraft
Extension Points	None
Alternate Flow	None

Scope of Concerns:
*Use Cases and
Inclusions*

Use Case 1 ---→ Use Case 1.1

Use Case 1.2   Use Case 2

Use Case → Actions

Functional Decomposition

FIGURE 5.7 Specification of Use Case description

The information in a Use Case description table is sufficient without further interpretation for mapping into an Activity diagram that shows the flow of actions for the actors and the system, which is depicted in Figure 5.8. The Activity diagram for the ATCS (as a black box) and a compliant aircraft was specified in Figure 5.3. The associated flow at the subsystem level of the ATCS represents the internal processes of the ATCS as a system. These were specified in Figure 5.4.

In the tutorial, Requirements and System Definition processes have been implemented in UML model specifications and transformation to create the Environmental

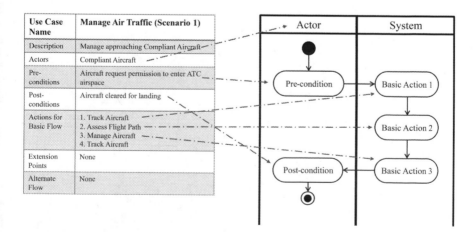

FIGURE 5.8 Mapping of Use Case description into basic flow of actions

and Behavioural Models that conform to the process, methods, and UML models prescribed in Chapter 4, Section 4.2 for Structured Analysis:

Use Cases → Included Use Cases (functional decomposition)
                → Use Case description
                → Activity diagram

For the Behavioural Model, the Activity diagram provides a graphical model that identifies what the flows are through the system and will be used to establish controls over the system flows. These models become the starting point for Structured Design and specification of the Essential Architecture.

6 Tutorial II
Air Traffic Control System Architecture

KEY CONCEPTS

Functional allocation
Concurrent definition
Reductionism and synthesis
Interoperation

The purpose of this tutorial is to illustrate the *practice* of essential activities in an Architecture Definition technical process. It shares a parallel structure with and gives insight into the second part of the practical case study in (Chapter 11, Section 11.2) but is not intended to solve it. The essential activities are aligned to key systems standards that were reviewed in Chapter 3. The practical activities of systems architecting demonstrated in this tutorial will develop models for:

Concurrent functional definition and allocation
Subsystem behaviour integration
Interoperation: System structure and subsystem interactions
Specification of the essential architecture for the system

The models in this tutorial will be developed in terms of new Unified Modeling Language (UML) diagrams (reviewed in detail in Chapter 9) that are based on Chapter 5 (Tutorial I). The essential architecture developed will be expressed as a holistic structure that exhibits the organisation and interoperation of system software and hardware. It will be traceable back to the essential system model from Tutorial I. UML diagrams provide sufficient precision and rigor to express the traditional process of system definition and decomposition in terms of model specification and transformation using a method of reductionism and synthesis. Definitions of the key terminology of systems and architecture for this tutorial are summarised in Chapter 3 and Annex A-3.

6.1 CONCURRENT FUNCTIONAL DEFINITION AND ALLOCATION

The concluding model in the first tutorial was a high-level Activity diagram (Figure 5.4) that depicted Air Traffic Control System (ATCS) system behaviour for the Compliant Aircraft scenario. This is the baseline for the system processes. The diagram followed from the

DOI: 10.1201/9781003213635-6

system level black box Activity diagram (Figure 5.3) with minimal discussion of the logical transition. In practice, an experienced system architect with sufficient domain knowledge can make a transformation between two models like this without showing details of how it was done. This might be appropriate for a simple system or problem such as in the tutorials but in complex systems or commercial applications, the details of the transformation and the rationale need to be exposed. In the first tutorial (Chapter 5), Exercise 5.3.1 raised the question of what is needed to complete the transformation from Figures 5.3 to 5.4. The first half of this second tutorial will be concerned with how to do this in a prescriptive way using the essential definition process and structured methods in Chapter 4.

It was proposed in that chapter to apply the essential system definition process to the elements in the first level system hierarchy in the same way that it is applied to the system itself. Structured methods were used to implement the process. It was also noted that the environment of each system element (i.e., subsystem) in the first-level hierarchy can include peer elements in the hierarchy as well as elements from the system-level operational environment. If the essential system definition process is applied in this way, Use Case diagrams must be specified for both the Air Traffic Radar (ATR) and Air Traffic Management System (ATMS) as subsystems of the ATCS. This involves a functional decomposition of the use cases of the ATCS (Figure 5.2), then specification of Use Case descriptions that will be the basis of the black box activity models of each of the subsystems. Figures 6.1 and 6.2 display Use Case diagrams for the ATR and ATMS based on Figure 5.2.

Recall that the use cases at the system level were defined concurrently across the actors (elements) in the environment. The functional decomposition process is similar to this but different in that it must be performed concurrently across the two subsystems (the ATR and the ATMS). The specification of subsystem use cases then becomes what is known in systems engineering as *functional allocation*, i.e., based on the use cases, the decomposed functionalities become associated with (assigned to) one or more of the subsystems. In a traditional systems engineering approach that does not employ Use Case diagrams, the subsystems would be identified and given

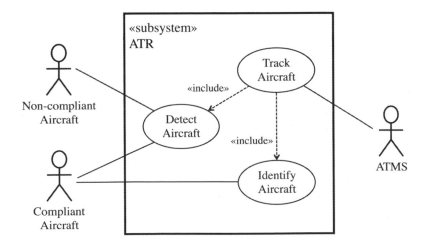

FIGURE 6.1   Initial Use Case diagram for the ATR

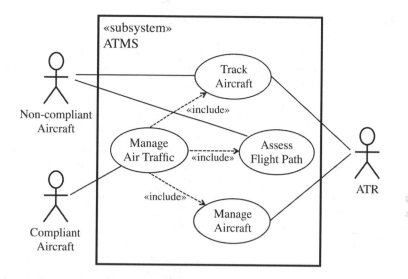

FIGURE 6.2 Initial Use Case diagram for the ATMS

names, a list of functions would be defined separately, then the functions would be assigned to the subsystems. The main difference between the traditional approach and a model-based approach is that interactions between elements would not be identified as part of the allocation process in the traditional approach. The model-based approach with Use Case diagrams, as depicted in the figures, explicitly models the peer elements in the environment of a sub-system as actors, e.g., the "ATMS" actor in the ATR Use Case diagram and the "ATR" actor in the ATMS Use Case diagram. New association relationships between these new actors and the sub-system use cases must then be specified. The functional allocations and the associations will need to be analysed before being used as the basis for the specification of sub-system behaviours.

Given that the two diagrams in Figures 6.1 and 6.2 have been specified independently of each other but share a use case, they provide a good example of the need for analysis in concurrent functional definition and allocation. When they are considered as a representation of the ATCS as a whole, it also becomes clear that analysis and normalisation will be needed. The first issue that is evident from the concurrent functional definition is that the ATR and ATMS have both been allocated the "Track Aircraft" functionality. A second less obvious issue arises from comparing the use cases in the figure with those in Figure 5.1 for the ATC. Specifically, clarification is needed to address the difference between the "Control Air Traffic" use case for the ATC and the "Manage Air Traffic" use case for the ATMS. Therefore, two key functionalities and their allocation must be addressed: "Track Aircraft" and "Manage Aircraft". Both domain knowledge and review with the stakeholders will be necessary to do this.

The issue of allocation of the "Track Aircraft" functionality will be addressed first, as the first tutorial has already provided a beginning. The diagram for the ATR in Figure 6.1 reflects that (i) the ATR is required to detect both types of aircraft and (ii) the ATCS must track both types of aircraft (as depicted in Figure 5.2). This tracking requirement at the system level was based on the narrative. However, based on domain knowledge of radar tracking, both detection and identification (based on the detections) are required from the ATR for tracking aircraft. Any aircraft that complies with flight regulations for entering the ATC airspace is required to operate a transponder. This is the only identification mechanism that has been specified at this point in the system definition of the ATCS. Non-compliant Aircraft likely will not have this equipment (or might not be using it if they do). Therefore, the ATR should not be expected to identify or track Non-compliant Aircraft. The detections though might be useful for cueing another type of sensor.

The allocation of tracking functionality leads to a further question: which subsystem of the ATCS should be allocated this functional requirement; or should it be shared, i.e., should it be a collaboration between subsystems? Depending on the allocation, external sensors or information sources might be involved with the detection and tracking of the Non-compliant Aircraft. Further domain knowledge of the technologies involved reinforces the importance of this question. Specifically, radar detection and tracking proceed in two high-level stages. The first is to illuminate the 'target' with radiofrequency energy then receive returned energy. Modern radars almost exclusively digitise the detections of this energy and process the digital data rather than the analogue signal. This type of signal processing of the received energy produces data (not tracks). The second stage is to process the data to produce tracks. The question is then whether the ATR should be doing both signal and data processing.

A top-down viewpoint on system and architecture definition must be reconciled with the technologies that will ultimately implement the system. This does *not* mean that a system design for implementation is needed before choices concerned with functionality can be made. Instead, what is needed is the efficient use of domain knowledge and the models that have been specified up to this point. For example, requirements concerned with radar location could and likely will lead to some degree of exposure of the ATR to weather. A casual survey of radars currently deployed at Class A and B airports would indicate that their radars are typically sitting at the highest possible point available and are not protected from the weather. It is questionable whether the computing equipment used for data processing should be exposed to uncontrollable environmental conditions. In conventional practice, the isolation of the computing equipment is accomplished by locating it in a separate facility. The control centre for ATC activities where the ATMS might be expected to reside has controlled conditions suitable for the equipment. Therefore, an initial architecture trade-off decision would be to consider the allocation of tracking functionalities to the ATMS and replace the "Track Aircraft" use case for the ATR with a "Support Track" functionality that specifically includes a "Report Data" use case. This would also support an ATCS design concept in which the ATMS accesses external sensors and information. (A trade-off decision like this can become a topic of discussion at the next stakeholder meeting.)

The second issue is perhaps easier to address but requires further domain knowledge. Specifically, the ATC Controller (ATCC) can be expected to assume final authority and responsibility for any directives made to the aircraft. Similarly, the pilot has final authority for the flight and safety of the aircraft; and can refuse to follow directives from the ATC. The Controller, therefore, only accesses the ATMS as a support service for flight path management.

Figures 6.3 and 6.4 depict the normalised subsystem Use Case diagrams based on a decision to allocate the "Track Aircraft" functionality to the ATMS. The role

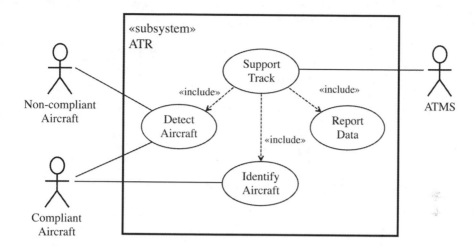

FIGURE 6.3   Normalisation of ATR Use Case diagram

of the ATCC in the ATMS Use Case diagram can also be a means to draw attention to the fact that this subsystem is providing a service and the control of aircraft is not a functionality that has been allocated to it. The use cases have now been specified in a way that is consistent and traceable to the system-level use cases. There is no conflict or contention between them at the subsystem level. They are also complete in the sense that they are sufficient to express the system level use cases. It should be noted that if a decision had been made for the ATR and ATMS to share the track functionality, further analysis and functional decomposition would have been required.

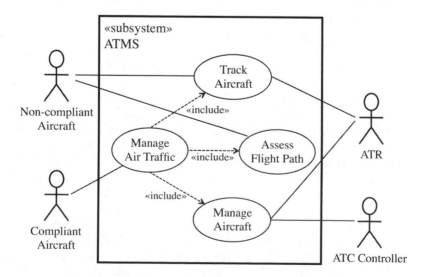

FIGURE 6.4   Normalisation of ATMS Use Case diagram

Exercise 6.1

1. Using Figure 6.3 as a reference, update the ATR Use Case description for the:
   a. Compliant Aircraft baseline scenario
   b. Non-compliant Aircraft alternative scenario
2. Using Figure 6.4 as a reference, update the ATMS Use Case description for the:
   a. Compliant Aircraft baseline scenario
   b. Non-compliant Aircraft alternative scenario

6.2   SUBSYSTEM BEHAVIOUR INTEGRATION

Now that the Use Case diagrams for the two subsystems of the ATCS have been normalised, the next step in the technical process is to apply transformation methods to create Activity diagrams capturing the detailed behaviour for each of the subsystems. In the transformation, behaviours of the actors in each sub-system environment will be included only as required to establish the basic flow of actions. The details of creating a Use Case description for each subsystem as an intermediate step will also be suppressed. The synthesis is based only on the final subsystem Activity diagrams for the baseline scenario (a Compliant Aircraft entering ATC airspace). To emphasise the subsystem viewpoint of each Activity diagram, its swim lane is shaded.

6.2.1   BEHAVIOUR OF INDIVIDUAL SUBSYSTEMS

In the transformation for the ATR, behaviours of the actors (in Figure 6.3) will be included only as required to establish the flow of actions of the ATR in the context of its environment. Thus, in the Activity diagram for the ATR, only the "Manage Aircraft" action will be included for the ATMS and only "Squawk" for the Compliant Aircraft.

The basic flow is modelled in the ATR Activity diagram in Figure 6.5 which has been developed from the Use Case diagram in Figure 6.3. The three lower level func-tionalities that support aircraft tracking have been organised into the basic flow based on domain knowledge acquired from radar engineers. The ATR first needs to detect the aircraft that is approaching the ATC airspace. This is achieved by sending out a pulse of radiofrequency energy to trigger the transponder on the target aircraft (i.e., to access its squawk function). The encoded signal returned from the transponder is received and processed by the ATR. This accomplishes aircraft identification. Other properties of the returned signal such as time between transmission and reception of signals and Doppler shift in the signal can be processed to provide aircraft data such as its range and speed of approach. These data are then reported to the ATMS for further analysis.

For the ATMS, based on the design decision to include aircraft tracking as part of the subsystem functionality, the basic flow of the ATCS in the Activity diagram in Figure 5.3 of the previous chapter can be used as a reference together with the Use Case diagram in Figure 6.4 to establish the subsystem basic flow depicted in Figure 6.6. The pre-condition for ATMS activity is "Activate ATMS" (initiated by the ATCC). The post-condition for completion is "Cleared for Landing". Once again,

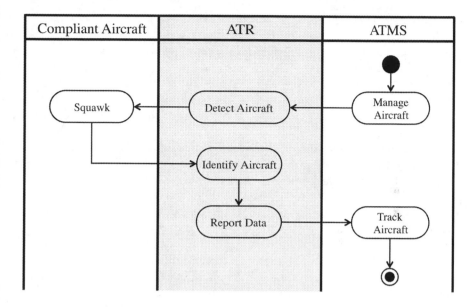

FIGURE 6.5   Subsystem Activity diagram: ATR

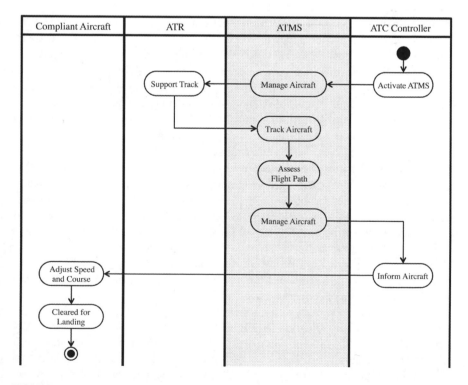

FIGURE 6.6   Subsystem Activity diagram: ATMS

only the required behavioural details of the actors in the environment of the subsystem have been used to create the Activity diagram. Note that the "Inform Aircraft" action in the ATCC swim lane provides for the resolution of the ownership of control of aircraft in the airspace. The ATMS is seen in different roles depending on which actor it interacts with. On the one hand, it is seen as a service that is activated by the ATCC and provides aircraft management recommendations to the ATCC; while on the other hand, it accesses data obtained by the ATR to track the approaching Compliant Aircraft.

In between these events in the Activity diagram, the tracking data obtained from the ATR is used to form a flight path of the aircraft that is then assessed against the tempo of other aircraft and potential risk of air-to-air collision (see the narrative provided in Chapter 5, Section 5.2). Depending on the outcome of the assessment, the ATMS will generate an appropriate recommendation that will then be used by ATC Controller to inform the aircraft pilots as to what to do next. The ATMS basic flow is completed when the Compliant Aircraft responds to the ATCC. The flow depicted in Figure 6.6 is also useful as an example of an interaction in a use case being indirect, i.e., the effect of the ATMS on a Compliant Aircraft in the flow of activities is through the ATCC. Comparing Figure 6.4 with Figure 6.6, it is clear that the ATMS does not manage aircraft through direct communication, but indirectly via the ATC Controller. This can become a discussion point for the next meeting with the stakeholders, which will clarify further the intended behaviours of ATMS and its interactions with the elements in its environment.

There are other important observations that can be made from this example. First, although the basic flows terminate at some point in the Activity diagram, for each Compliant Aircraft entering the ATC airspace, the behaviour of the ATR and ATMS will continue by repeating their basic flows until it is cleared for landing. Modelling this requires the use of Decision and Merge nodes as explained in Chapter 8. This is a type of control structure. It is not intended to be modelled in the baseline scenario.

Next, when architecting the behaviour of individual subsystems, their environment consists of not only interacting elements in the system operational environment, e.g., the Compliant Aircraft, but also the other subsystems. Within the context of subsystem behaviour, these elements are modelled as Actors. As such, the specification of their behaviours remains at the (black box) system level as if the subsystem of interest has no knowledge of the processes within the elements of its environment. For example, as in Figure 6.5, the ATR does not see the detailed behaviours of the ATMS which are modelled in Figure 6.6. Similarly, ATMS in Figure 6.6 does not see the detailed behaviours of the ATR which are modelled in Figure 6.5. This may seem to be overcomplicating the modelling processes for a simple system composed of only two subsystems. In fact though, this approach simplifies the modelling task by managing the complexity through the separation of concerns. This is obvious in practical engineering problems where a subsystem can interact with multiple other subsystems.

Finally, the Activity diagram in Figure 6.6 captures a flow from the "ATC Controller" to the "Compliant Aircraft". This flow may seem to contradict minimalism in the diagram because it represents a possible interaction between two elements in the environment of the ATMS that do not directly involve it. In fact, this flow is necessary and exposes an interesting architectural decision as to how the ATMS manages air traffic. This was modelled by the association between the ATMS "Manage Air Traffic" use case and the "Compliant Aircraft" actor as shown in Figure 6.4. Recall from the comparison of Figure 6.4 with Figure 6.6, it was concluded that the ATMS does not manage the aircraft through direct communication, but indirectly via the ATC Controller.

6.2.2 BEHAVIOUR OF THE INTEGRATED SUBSYSTEMS

This is a pivotal point in the modelling of the system. It is important at this point to understand that traditional system definition and decomposition employ a reductionist methodology. This is something that has been used extensively in science and mathematics. In fact, Newton was one of the early pioneers of reductionism. Systems engineering on the other hand has traditionally been practiced without the mathematical foundation enjoyed by science. UML diagrams provide sufficient precision and rigor to implement system definition and decomposition in terms of model specification and transformation. This can support a rigorous method of reductionism and synthesis suitable for engineering. The graphical notation of UML is very useful because the succinct terms in the nodes and the syntax of connecting the nodes support a precise approach to maintaining semantics and relations within the decomposition of the system, whether functional or in term of elements.

The easy part of reductionism is decomposition of a whole into parts. It is more difficult to synthesise the parts back into an integrated whole. This tutorial has already begun synthesis of behaviours which can be seen in the activity models developed in Figure 6.5 for the ATR and Figure 6.6 for the ATMS. Any method of synthesis is challenged by the complexity of integrating the multiple interrelated parts back into a whole without violating their meaning and relations. This explains why the simpler model associated with the radar behaviour (process) was chosen first. The insistence on minimalism in this choice as well as in the specification of the models is key to managing the complexity. Recall for example, the suppression of detail in the behaviour of the actors in the system environment of ATR. Only one action of each actor was specified: "Squawk" for "Compliant Aircraft" and "Manage Aircraft" for the "ATMS". Furthermore, the modular structure of the subsystems resulted in the sharing of only one actor, the "Compliant Aircraft". The ATR and ATMS do not share any behaviours of this actor, which further serves to simplify the synthesis.

It is therefore straightforward to combine the Activity diagrams in Figures 6.5 and 6.6 into the Activity diagram depicted in Figure 6.7, which provides an integrated behaviour model for the system at the first-level hierarchy. The challenges of complexity in synthesis have been reduced.

In the process of synthesis, new actions should not be introduced in the swim lanes (partitions defined by the subsystems and actors). While putting the ATR partition and the ATMS partition side-by-side, the inner flows of the subsystems are inherited from the previous two figures, but the crossovers between partitions require careful consideration to appropriately account for the intended order of actions at this level of the hierarchy. For example, in the ATMS Activity diagram, the ATR plays the role of an actor and its behaviour has been simplified to the single action "Support Track". However, in the ATR Activity diagram, the ATMS plays the role of an actor

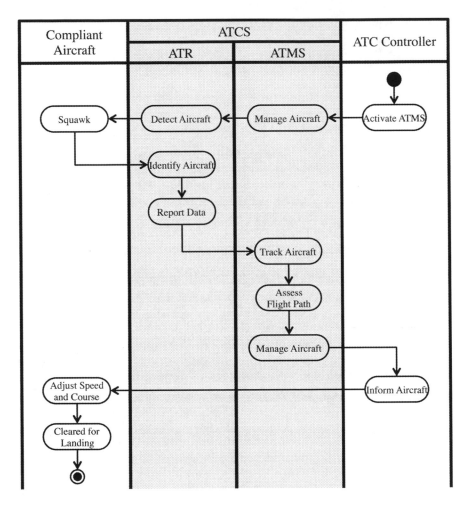

FIGURE 6.7   Synthesised ATCS Activity diagram

that enters the ATR swim lane at the "Detect Aircraft" node and returns to the ATMS "Manage Aircraft" node. These flows have been normalised in the integrated Activity diagram in Figure 6.7 where the flow through the ATR returns to the ATMS "Track Aircraft" node. It should also be noted that the inclusion of the behaviour of the elements in the ATCS operational environment, e.g., Compliant Aircraft, now exposes details that have been suppressed in the subsystem models.

Exercise 6.2

1. Using Figure 6.5 and the Use Case description for Non-compliant Aircraft (Exercise 6.1.1b) as a reference, specify the alternative scenario ATR subsystem Activity diagram.

2. Using Figure 6.6 and the Use Case description for Non-compliant Aircraft (Exercise 6.1.2b) as a reference, specify the alternative scenario ATMS subsystem Activity diagram.
3. Using Figure 6.7 and the previous two exercises as a reference, specify a synthesised ATCS Activity diagram for the Non-compliant Aircraft alternative scenario.

6.3   SYSTEM STRUCTURE MODEL

In the context of structured analysis and design, it is important to keep in mind that modelling is still at the class level of abstraction. Structured Design as presented in Figure 4.7 of Chapter 4 is a method to implement an architecture definition process. Classes are used for the specification of types of objects rather than the objects themselves. This is an important distinction that helps to prevent solutions engineering instead of systems engineering.

The first model specification and transformation in the relational framework depicted in Figure 4.7 proceeds in two steps. A class has been named for each of the two ATCS subsystems in the Activity diagram depicted in Figure 6.7. The nodes in the activity flow are actionable use cases for the ATR and ATMS which were specified in Figures 6.3 and 6.4, respectively. The associated classes must have defined operations that enable the actions in the Activity diagram. This type of process is similar to the functional allocation of functionalities to system elements (subsystems). It is appropriate to also assign requisite operations to the classes that represent the actors, in this case, the "ATCC" and the "Compliant Aircraft". The five classes depicted in Figure 6.8 have done this based on a variety of sources to include domain knowledge.

The second step in this first model specification and transformation is to identify the intended relations between the classes. There are two types in the figure. One is decomposition and the other is message exchanges between the classes. The aggregation and composition relations for the ATCS regarding the ATR and ATMS subsystems are based on design decisions. The radar subsystem is acknowledged as pre-existing in the current ATC service but the ATMS is an integral part of the developer proposal for the system solution. Exchange relations are based on domain knowledge.

The Class diagram must also include and be consistent with every relation and interaction derived from the two Use Case diagrams in Figures 6.3 and 6.4. For example, the ATCS as a system interacts with the actors in its operational environment (i.e., the ATC Controller, Compliant Aircraft) through its subsystems (i.e., the ATR, ATMS). The Structure Model specified by the Class diagram, therefore, exhibits the elements of the system and its environment together with the relations between the elements.

Exercise 6.3

Draw the ATCS Class diagram for the Non-compliant Aircraft alternative scenario.

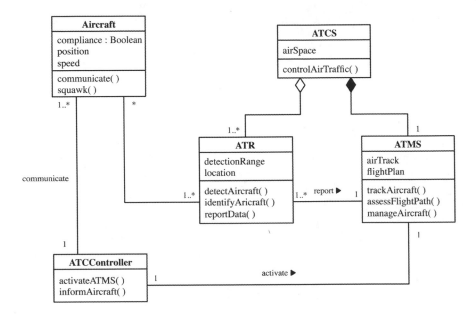

FIGURE 6.8   ATCS Class diagram

6.4   SUBSYSTEM INTERACTION MODEL

The Interaction Model is the final specification in the Architecture Definition process as depicted by the framework in Figure 4.7. The Sequence diagram depicted in Figure 6.9 for the ATCS baseline scenario (Compliant Aircraft) is a UML Interaction diagram. It is not per se 'a model of the ATCS architecture'. Rather, it synthesises a constrained interrelational structure as noted in the Chapter 4, Section 4.5 discussion of Figure 4.11. Every model element from the Activity diagram in Figure 6.7 and the Class diagram in Figure 6.8 together with their semantics and relations must be preserved in the synthesis process of the underlying interrelational structure. Specific details are provided in the discussion of Figure 6.12 in Section 6.6.2. The structure and properties synthesised and normalised from the Activity and Class diagrams are expressed in the Sequence diagram. They form a comprehensive holistic architecture of the ATCS. The diagram in Figure 6.9 is the first *object* level model of the system that has been developed up to this point. As noted in the discussion of the subsystem structure model, the representations of the system up to this point have been in terms of classes and properties.

Interoperation is understood at the object level of system modelling and depends heavily on Interface Definition. This will be investigated in the next and final tutorial (Chapter 7). It is clear though in the Sequence diagram depicted in Figure 6.9 that the ATR and ATMS collectively have three interfaces. The object-level model represented by the Sequence diagram together with the other models can be sufficient for enabling Design Definition, or even implementation in the same sense that computer code is intended to be deployed from the software architecture of a system in the OMG Model Driven Architecture (refer to Chapter 1).

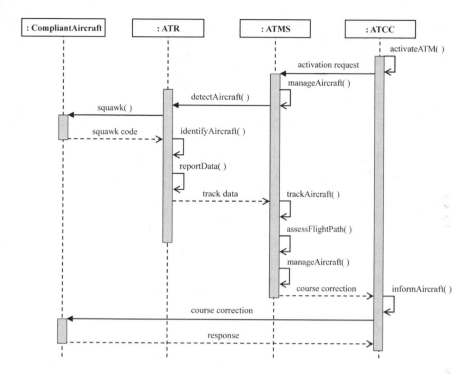

FIGURE 6.9   Sequence diagram: ATCS baseline scenario

Exercise 6.4

Draw the Sequence diagram for the ATCS Non-compliant Aircraft alternative scenario.

6.5   SPECIFICATION OF THE ESSENTIAL ARCHITECTURE

At this stage of the engineering project to deliver an ATCS to the acquiring stake-holder, an agreement has been reached that the proposed system solution will have increased functionality beyond that of the current operating systems in the ATC service. Specifically, the ATCS proposed will mediate operations between the airport ATCC and the Compliant Aircraft that enter the ATC air space. In this second meeting with the customer, the purpose is to review and seek agreement on the concepts of and new functionalities for the system proposed by the developer.

The architect has developed a technical package that is suitable for the review with the stakeholders. The central artefact in the package is the essential system architecture for the ATCS, which should be traceable back through the models given to the customer in the first meeting. The complete package would be comprised of all the diagrams in the figures in this section along with those in the exercises. As with the first stakeholder meeting, the higher level UML graphical models would be used in an executive-level presentation whilst the more detailed models would be intended for the technical team of the stakeholder. These can also be used to address specific questions of detail that might arise during the executive presentation. As mentioned in the first tutorial (Chapter 5), the architect ideally would be in a position to argue that the technical package has been produced through a qualified model-based systems engineering process.

The executive summary of the package could be presented as a checklist that indicates project status and points out key items for the next step in design and development. The example below follows a format similar to the executive summary in the first tutorial (Chapter 5, Section 5.5). A favourable outcome of the meeting for the developer would be a decision to proceed to the next stage. The detailed technical package could be presented in a side meeting with the technical specialists from the acquiring stakeholder.

- An essential system architecture has been completed for the ATCS:
  - Two system elements (ATR, ATMS) with specified functionality and behaviour
  - The modular design has minimised the number of interfaces needed (only 3)
- ATCS subsystem functional definition proposed for the ATMS includes:
  - Track Aircraft, Assess Flight Paths, Manage Aircraft
  - Functionality from both the ATCC and ATR are being integrated into the ATMS
  - ATCS processes are specified based on UML Activity & Sequence diagrams
- Behaviours of the ATC and ATCS systems have been specified in two key scenarios:
  - Compliant Aircraft (the baseline scenario)
  - Non-compliant Aircraft
- Performance and non-functional requirements will be analysed for the next meeting

Traceability to the narrative has been ensured by a state-of-the-art model specification and transformation process. The architecture team is ready to seek agreement between stakeholders and engineering to proceed to design readiness and implementation *but need agreement on the ATR concept and the significance of Non-compliant Aircraft in terms of customer requirements.*

- Technical details, requirements, and domain knowledge have been expressed in UML to facilitate precision and communication, and deployment of the essential architecture to software development. These have been produced by a rigorous model-based systems engineering technical process that is aligned to modern standards.

6.6   SUMMARY OF MODEL SPECIFICATIONS AND TRANSFORMATION

The chapter summary is comprised of two parts. The first is a refinement of the system concept provided in the stakeholder narrative that is briefly summarised from the viewpoint of the system developer. The second is a specification of the types of models used in the chapter and their transformation.

6.6.1 REFINEMENT OF THE SYSTEM CONCEPT

The ATC service proposed by the developer will be implemented by an ATMS and an ATR that will mediate operations between the airport ATCC and the Compliant Aircraft that enter the ATC air space. These ATMS and ATR elements form the first-level hierarchical decomposition of the ATCS. Two key actors were also specified in the essential architecture of the ATCS: Air Traffic (comprised of Compliant and Non-compliant Aircraft) and the Airport ATCC that provides direct control of the aircraft. (These can be thought of as clients of the ATCS service.) These four elements define the system boundary of the proposed service.

The system developer has provided the high-level technical details for integrating ATC track management functionalities into the ATCS. The ATC service will be improved by reducing the workload on the ATCC. The primary focus of the analysis at this stage of the proposal has been on the specification of the ATCS subsystems (i.e., the ATMS and ATR) and system interoperability, and identification of interfaces which trace to the key functionalities defined for the ATC service: to track and manage the aircraft, and to assess their flight paths.

At the next stakeholder meeting, detailed analysis will be presented for design decisions. This will include the previously identified performance metrics for the functionalities such as the detection range of the radar and tempo of the airport. The customer still has made no indication as to whether the ATR needs to be upgraded or replaced. This will also be a subject of the stakeholder meeting.

6.6.2 MODEL TYPES AND TRANSFORMATION

Chapter 4 introduced processes and structured methods for Architecture Definition. Two types of models were specified in terms of UML: Class and Sequence diagrams. The Use Case and Activity diagrams from the previous work on System Definition are also used extensively in Architecture Definition. The tutorial in this chapter has implemented the Framework for Structured Design that was depicted in Figure 4.7. The model specifications and transformation in this second tutorial will be summarised using the results presented in Figures 6.1–6.9. Specifically, they are organised around

- Concurrent functional definition and allocation
- System behaviour integration
- Specification of system structure model and subsystem interactions
- Synthesis of the essential architecture

The architecture can also be used for an initial identification of system interfaces.

The model specifications and transformation in the relational framework for Structured Design depicted in Figure 4.7 proceeds in two steps. The starting point is the behavioural model from the System Definition process that was specified using Structured Analysis. An (implementation free) Activity diagram was defined at the system level, as were basic flow models for each element (subsystem) at the first level of hierarchy. At this point, a class has been identified for each of the subsystems in the Activity diagram depicted in the Essential Model. The nodes in the basic

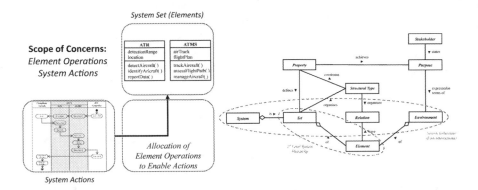

FIGURE 6.10 Functional allocation to the system set

activity flows are actionable use cases for the subsystems and each subsystem is represented by a class. The association must also specify defined operations that enable the actions in the Activity diagram. However, the behaviours of the subsystems that have been specified concurrently should first be integrated. This is preparatory work that 'initialises' the Structured Design process in Figure 4.7 and was the subject of Sections 6.1 and 6.2.

The process then proceeds as depicted in Figure 6.10. It is similar to the allocation of functionalities to system elements (subsystems) but it must be done from the integrated behaviour model. Thus, the functionalities allocated at the system level and modelled by use cases are decomposed to the subsystem level and then associated with defined operations in the subsystem classes. This is how the system element functions are specified in the relational framework for Structured Design using functional allocation. It is also appropriate to assign requisite operations to the classes that represent the actors (in the environment).

Figure 6.11 illustrates the transformation of the ATCS subsystems into a system structure that was used in Section 6.3 of this chapter to concurrently model the system and the key actors that interact with it. The logical diagram on the right side of

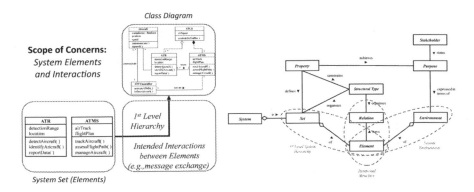

FIGURE 6.11 Specification of system structure model

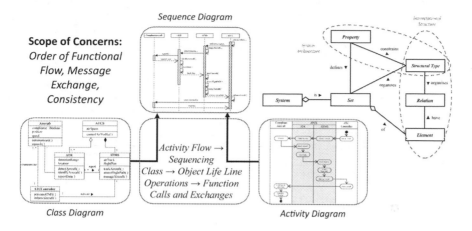

FIGURE 6.12 Synthesis of essential architecture

the figure depicts the system set and elements of the environment both of which are represented as UML Classes in the transformation. The actions of the previously developed Activity diagram and the system hierarchy are the basis for specifying the intended message exchanges between the classes. (Recall that the class operations enable the actions.)

The structure and properties synthesised and normalised from the Activity and Class diagrams are expressed in the Sequence diagram. They form a comprehensive holistic architecture of the system. (Refer to Chapter 4, Section 4.3.1 and Section 6.4 of this chapter.) The synthesis process for the ATCS is depicted in Figure 6.12. The model must preserve the sequencing of activity flow of system elements, the interactions (e.g., exchanges) between elements, and the control structure that governs the flow. In terms of UML model elements, the model transformations are realised by the following associations: Class → Objects → Lifelines; Operations → Function Calls. The resulting *Essential Architecture* defines an interrelational structure for the system hardware and software, as depicted by the logical diagram in the upper right of the figure. The system elements can be refined to enable Design Definition or to specify an *Implementation Model* depending on whether sufficient detail has been specified or not.

Figure 6.13 provides details of the synthesis for creating the ATCS Sequence diagram. The actions in the Activity diagram in Figure 6.7 of the tutorial have been associated with the class operations in Figure 6.8. The associated flow at the subsystem level of the ATCS represents the internal processes of the ATCS as a system. This flow is now seen to progress downwards through the lifelines of the Sequence diagram.

The Sequence diagram in Figure 6.9 can provide the level of detail necessary to begin behavioural Interface Definition for the three interfaces that were identified for the interoperation of the ATR and ATMS. These will be defined in the next and final tutorial (Chapter 7). Specifying models for the interfaces is part of the detailed Structured Design specification for implementation of the Essential Architecture. The hardware and software elements are represented by the objects of the Sequence

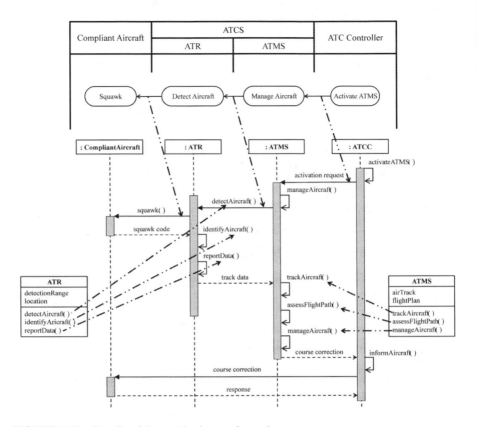

FIGURE 6.13   Details of the synthesis transformation

diagram. The model specifications and transformation used in this second tutorial complete the first recursion of essential System and Architecture Definition processes for the ATCS.

If the details can be agreed up by all the stakeholders and are sufficient to buy or build the hardware and software elements of the system, then Design Definition (as specified in ISO/IEC/IEEE 15288:2015) has been completed. Otherwise, iteration of the details as part of a refinement process may need to be conducted with both the customer and developer until the final agreement is reached. If refinement through an iterative process does not result in a sufficient level of detail and agreement for completing Design Definition, then System and Architecture Definition must further be applied recursively (and iteratively) until a sufficient level of detail and agreement has been achieved. The refinement process and final system design specification of the ATCS will be the subject of the next tutorial.

7 Tutorial III
Air Traffic Control System Final Solution

KEY CONCEPTS

Modelling and analysis
Transformation of structure
Interfaces
System specifications

The purpose of this last tutorial is to illustrate the *practice* of essential activities in both System Definition and Architecture Definition in an iterative manner to derive a final solution that includes software, hardware, and interface specification, and assurance of the architecture through initial test case definition. It shares a parallel structure with and gives insight into the third part of the practical case study (Chapter 11, Section 11.3) but is not intended to solve it.

The practical activities of system architecting and design demonstrated in this tutorial will develop models for:

Iterative Architecture Development
Interface Definition
Architecture Analysis
Specification of the Implementation Model

New models in this tutorial will be developed in terms of the Unified Modeling Language (UML) diagrams in the previous chapters that are complemented by new Systems Modeling Language (SysML) diagrams (reviewed in detail in Chapter 10) to deliver the final architecture and system design of the solution. The tutorial will be concluded with a Summary of Model Specifications and Transformation as usual, that is further followed by a reflection on the Art of System Architecting and Design for the whole tutorial case study part.

In practical application, the outcome of the four activities for the final solution should be readiness to 'buy or build' the components of the proposed system. To buy a component can require less detail in the Implementation Model than to build it. A build decision can and likely will be followed by further System and Architecture Definition involving recursion (second-level decomposition) and iterative analysis as part of the Design Definition of the component.

DOI: 10.1201/9781003213635-7

This is a good point in the tutorials to remind the reader and especially the student that the intent of the material in this chapter is instructive and not to specify a detailed implementation model that is sufficient for building or buying the components of a real system, but rather to give the tutorial case study a sense of realism.

7.1 ITERATIVE ARCHITECTURE DEVELOPMENT

The ATCS system architecture developed in the second tutorial has led to an initial software specification for the ATCS. However, this is not quite complete for the ATCS software in terms of capturing alternative behaviours to the baseline scenarios, for example, the need for occasionally dealing with Non-compliant Aircraft (NCA). The scenario for NCA so far has only been so far modelled in a set of separate diagrams specified in the exercises in the previous two tutorials. Essentially, these specify a behaviour (process) for the ATCS to advise the ATCC on how to control Compliant Aircraft to deal with NCA. The specified behaviours themselves are appropriate, but there is an issue for the integration of the behaviours: how does the software know an incoming aircraft is compliant or not such that it can trigger the correct corresponding behaviour?

This type of issue can emerge during integration activities. It resulted from a 'God's eye view' that was taken in the initial system and architecture definition process. Following such a viewpoint is natural in the process of transforming stakeholder needs to system and architecture specifications. Nonetheless, such problems must eventually be rectified before claiming readiness for design or implementation. This part of the tutorial will demonstrate an iterative development of the architecture to address issues like this, thereby evolving the system design into its final form.

7.1.1 Alternative Behaviour

The first step in early-stage integration of behaviours is achieved by appropriately capturing the alternative behaviours in the system-level Use Case diagram. This is accomplished through extended use cases and functional decomposition to ensure that alternative behaviour is introduced at the correct 'place', which will be specified as an extension point.

Based on the Use Case diagram in Figure 6.3 where both actors, "Compliant Aircraft" and "Non-compliant Aircraft" are associated with the "Detect Aircraft" use case, it is possible to derive a functional requirement that states: for any aircraft approaching to the ATC airspace, the ATCS shall be able to declare whether the aircraft is compliant or non-compliant following the initial aircraft detection. This requirement then enables the decomposition of the "Detect Aircraft" functionality to capture the high-level description of the detection of a Compliant Aircraft as the baseline scenario, as modelled in the Use Case diagram in Figure 6.3; and additionally, the high-level description of the detection of a Non-compliant Aircraft as an alternative scenario, which is now modelled in the iterated Use Case diagram depicted in Figure 7.1.

In this diagram, the decomposition of the "Detect Aircraft" use case is modelled by an extending use case, "Declare Aircraft to be NCA", where NCA stands for Non-compliant Aircraft. As previously discussed, technically, the detection of an aircraft is achieved by the ATR transmitting a microwave beam to trigger the transponder mounted on a Compliant Aircraft which will then transmit a microwave beam back to the ATR with a squawk code that allows the ATR to identify the aircraft. In the

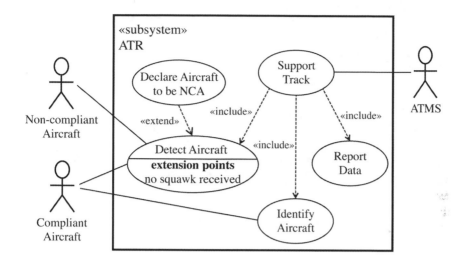

FIGURE 7.1   Iterated ATR Use Case diagram

absence of the transponder triggering event, the ATR would still receive reflected energy, but would only acknowledge the detection of an unidentified flying object. Utilising this domain knowledge in how radar detection works, the condition for which the extended use case is executed can then be modelled by an extension point in the "Detect Aircraft" use case, specified as "no squawk received", conforming to the syntax and semantic rules of UML.

Exercise 7.1.1: A Faulty Transponder

At this point, the iteration in the Use Case diagram does not consider the further splitting of the NCA scenario into an actual NCA and a supposedly compliant and identifiable aircraft with a faulty transponder. Further decompose the iterated ATR Use Case diagram to capture the alternative scenario where a Compliant Aircraft has a faulty transponder that fails to squawk.

7.1.2   CONTROL STRUCTURE

With the iterated Use Case diagram, the system-level behaviour as modelled in the set of UML Activity diagrams should reflect the iteration consistently. This is achieved by following the set of semantic transformation processes exactly as introduced and applied in the previous tutorials, as depicted in Figures 5.7 and 5.8.

As an intermediate step towards the iterative rectification of the Activity diagram in Figure 6.7, a Use Case description for the ATR "Support Track" use case is first created and presented in Table 7.1. Due to the involvement of multiple actors, the descriptions of actions are made slightly more comprehensive with inclusion of the involved actors. The construction of this table followed the same principle used in the construction of Table 5.1 in the first tutorial. The resultant table now contains an alternate flow in addition to the basic flow with the extension point reflecting the one modelled in the Use Case diagram in Figure 7.1. More importantly, the table captures when this extension

TABLE 7.1

Use Case Description: Support Track

Use Case Name	Support Track (Integrated Scenario)
Description	ATR supports tracking of an approaching aircraft
Actors	Aircraft (Compliant or Non-compliant)
Pre-conditions	Aircraft approaching the ATC airspace and ATR is activated by ATMS.
Post-conditions	ATMS acknowledges the type of the aircraft and tracks it
Actions for Basic Flow	1. ATR Detects Aircraft
	2a. ATR Receives Squawk from Aircraft
	3. ATR Identifies Aircraft
	4. ATR Reports Detection Data to ATMS
Extension Points	2b. ATR Does not receive Squawk from Aircraft
Actions for Alternate Flow	3. ATR Declares Aircraft to be NCA
	4. ATR Reports Detection Result to ATMS

point occurs in the basic flow through the usage of a numbering system. In this case, "2b" is used to indicate that the extension would occur at the second step in the basic flow and substitute the basic flow with the alternate flow subsequently.

This Use Case description can then be elaborated into an iterated Activity diagram, which is depicted in Figure 7.2. The basic flow and the alternate flow are integrated using a decision and merge nodes pair, forming an alternate flow control structure, also known as *exclusive choice*. In particular, following the execution of the "Detect Aircraft" action,

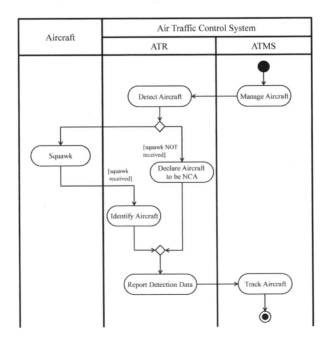

FIGURE 7.2 Iterated ATCS Activity diagram

the flow takes onto either of the two modelled paths. The condition for which path is taken is specified by the corresponding guard content, "squawk received" and "squawk NOT received", respectively. The two paths merge back just before the execution of "Report Detection Data", i.e., Step 4 in both of the basic flow and the alternate flow as depicted in Table 7.1. The traceability of the actions in the iterated Activity diagram to the defined actions, pre-condition, and post-condition in the Use Case description is self-evident.

Exercise 7.1.2: A Faulty Transponder Continued

Following the same approach presented in Section 7.1.2 and using Table 7.1 and Figure 7.2 as references, transform the answer to Exercise 7.1.1 into an Activity diagram capturing all three scenarios: baseline scenario for Compliant Aircraft, alternative scenario one for NCA, and alternative scenario two for Compliant Aircraft with a faulty transponder.

7.1.3 INTEGRATION OF SCENARIOS AND BEHAVIOURS

In this last step of iterative architecture development, the Sequence diagram is updated according to the revision of the Activity diagram, following the synthesis approach presented in Figure 6.12. According to this transformation, the Class diagram of the system needs to be updated to reflect the new actions introduced, in this case, "Declare Aircraft to be NCA". This action will be allocated to the "ATR" class based on the Activity diagram in Figure 7.3. It is specified as a new operation, "declareNonCompliance ( )", added to the "ATR" class; and it is the only revision to the diagram. Hence it will not be reproduced here. The expected output of this operation is the instantiation of the "Aircraft" class to create a new object "nonCompliantAircraft#: Aircraft", i.e., an object typed by the "Aircraft" class with the object name "nonCompliantAircraft#" where # is an ID number given

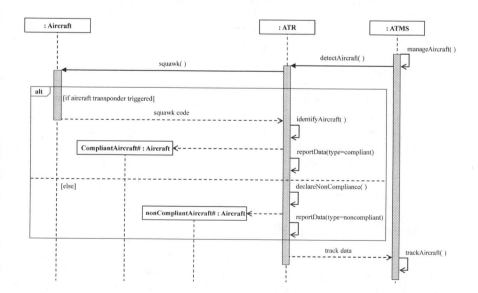

FIGURE 7.3 Iterated ATCS Sequence diagram: Integration of scenarios

to the NCA. The object has the Boolean attribute "compliance" taking a "False" value. For simplicity, the baseline scenario is not modified to have a symmetric operation "del-careCompliance ( )". Rather, an architectural decision is made here where the expected output of the "identifyAircraft ( )" is the instantiation of the "Aircraft" Class to create a new object, "compliantAircraft#: Aircraft". The creations of the objects in different scenarios are reflected in the Sequence diagram shown in Figure 7.3. The combined fragments with an "alt" interaction operator are used to properly transform the alternate flow control structure modelled in the Activity diagram into an 'if… then; else' exclusive choice control structure.

It is worth noting that for consistency, the returned squawk code is included in the first fragment of the combined fragments (see Figure 7.2 for reference where "Squawk" action happens after the decision node). However, from a software design and implementation perspective, this is not intuitive as the software owned by the ATCS does not know that if the aircraft transponder is triggered or not before the squawk code is received. In other words, the condition statement "aircraft transponder triggered" cannot be directly translated into a checkable code. As concluded in Chapter 6, if this is a sufficient software specification for a supplier (stakeholder) to provide the software solution, the implementation model for the ATCS will not be further decomposed. One can also imagine that this software specification then becomes the requirement for the software architects of the supplier to develop their software solution. If the stakeholders do not agree with this level of detail, this would require the architecture to be further recursively and iteratively developed to address the issue. A possible approach might be to set a response time window for the returning squawk code to arrive. If the code does not arrive within the window, it is identified as a Non-compliant Aircraft.

Exercise 7.1.3: Completion of the ATCS Implementation Model

1. Using Figure 6.9 as a reference, complete the Sequence diagram in Figure 7.3 in terms of interaction between ATCS elements with the ATCC and how ATCC interacts with Compliant and Non-compliant Aircraft.
2. Using the answers of the previous two exercises as the basis, further revise the Sequence diagram to include the behaviour of the ATCS for scenario where a Compliant Aircraft has a faulty transponder.

Finally, the structure and properties expressed in the UML diagrams up to this point in the tutorial form a comprehensive architecture for the deployment of hardware and software. The implementation model organised around the Sequence diagram (together with the results from the above exercises) specifies a final solution for the ATCS design. This is intended for the purpose of instruction only in the tutorial. This is a reasonable point to stop. The models can be used as the basis for the final software specification. The continuation of the tutorial past this point will investigate other aspects of the solution.

7.2   INTERFACE DEFINITION

This part of the tutorial demonstrates the identification and modelling of system interfaces. Specifically, the demonstration utilises the models of the baseline (Compliant Aircraft) scenario developed in Chapter 6 for the definition of system interfaces with respect to Compliant Aircraft.

7.2.1 INTERFACE IDENTIFICATION

The identification of system interfaces typically begins with interactions identified in the Use Case diagram for the system. The ATCS interacts with the aircraft and the ATC controller (the ATCC) but the exchanges with these two actors are at the subsystem level (i.e., the ATR and ATMS). Figure 7.4 illustrates how use cases can be annotated to identify candidate interfaces at this level of the system definition. Initially, three interfaces for the ATMS are identified based on four interactions (1, 2, 5, and 6) with the three key actors ("ATR", "ATCC", and "Compliant Aircraft") in its subsystem environment. Similarly, two interfaces for the ATR are identified based on three interactions (2, 3, and 4) with the two key actors ("ATMS" and "Compliant Aircraft") in its subsystem environment. The interactions have been numbered for clarity and traceability but in general, no order needs to be implied or intended by the numbering. It is important to remember though that Use Case diagrams only provide information about interactions (which may be direct or indirect) and do not necessarily imply exchanges between the objects of the classes associated with the interactions.

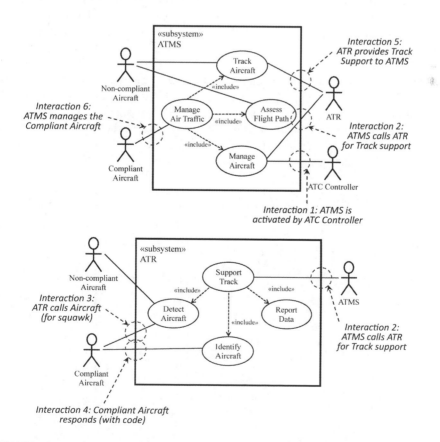

FIGURE 7.4   Interface identification with Use Case diagrams

Not all of the interfaces identified are valid. The next step is to confirm the identification by tracing to the Sequence diagram in Figure 6.9. The annotations in Figure 7.5 expose the details of the interactions, which have been modelled as an exchange of messages. As annotated in Figure 7.5, two of the three identified interfaces for the ATMS are confirmed by exchanges in the Sequence diagram. These are an ATMS–ATCC interface and an ATR–ATMS interface. On the other hand, the interaction between the ATMS and the Compliant Aircraft (Interaction 6) is an indirect interaction that has been achieved through the communication between ATCC and the aircraft. Also, both of the interfaces identified for ATR are confirmed by the Sequence diagram: an ATR–CA interface and an ATR–ATMS interface, in which the latter interface is also confirming the consistency with interface identification for ATMS.

In addition to confirming results from Interface Identification, the Sequence diagram in Figure 7.5 also provides the level of detail necessary to begin behavioural Interface Definition for both the ATR and the ATMS. In the interface between the ATR and Compliant Aircraft, a function call is made that is intended to trigger the transponder. The expected response is a reply with the "squawk code" that has been provided to the aircraft by the ATCC. This is used by the ATR to identify the aircraft and report data for tracking support, which is part of the exchange in the interface between the ATR and ATMS. It should be noted that this code must be passed from the ATCC to the ATMS then to the ATR as part of the "manageAircraft ( )" function call.

The complexity of the interrelations as captured in the Sequence diagram is now becoming evident. In the behavioural definition of the interface between the ATMS and the ATCC, the activation request must include the squawk code. After messaging the ATR, a chain of events must also occur before the ATMS can return a recommended course correction to the ATCC, thereby completing the "behavioural contract" of the interface.

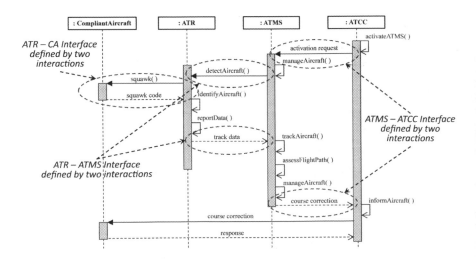

FIGURE 7.5 Interface definition with Sequence diagram

7.2.2 INTERFACE MODELLING

Model specifications for interfaces are one of the starting points for detailed Structured Design specifications of a hardware and software implementation of the system models bound together by the Essential Architecture. To model the interfaces as identified and defined previously, SysML Block Definition and Internal Block diagrams are considered more appropriate than using UML diagrams. This is because a complex interface may concern not only a software solution for information flow, but also hardware for energy and material flow. SysML ItemFlow, as explained in Chapter 10, provides a means to model these flows while SysML Blocks complement the flows to form part of the interrelational structure of the system elements.

The UML Classes that represent the ATR and ATMS are now re-interpreted as SysML blocks for the sake of modelling physical interfaces. Hence, the attributes and operations previously specified are not being redundantly modelled. Other means of combining the usage of SysML Blocks and UML Classes can be adopted, as long as there is a clear separation of concerns in which model traceability is maintained whilst the model complexity is managed.

As shown in the Block Definition diagram in Figure 7.6, the identified interfaces are modelled as a pair of ports, each owned by one of the blocks involved. For example, the ATR-CA interface is modelled as two full ports, an ": antenna" port which is owned by the "Air Traffic Radar" block, and a ": transponder" port which is owned

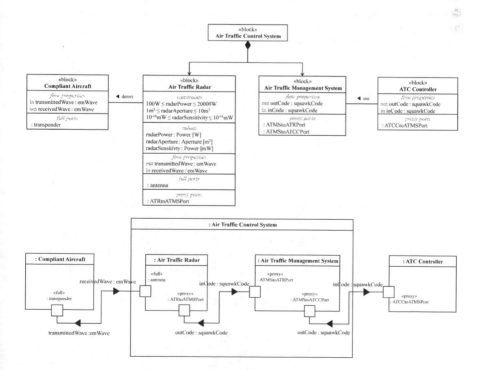

FIGURE 7.6   Interface modelling with Block Definition diagram and Internal Block diagram

by the "Compliant Aircraft" port. The behaviour of the interface is further modelled by the item that flows between the ports, e.g., the electromagnetic wave that propagates between the antenna and the transponder. This is modelled by the ": emWave" flow properties. The blocks are then instantiated in the Internal Block diagram to provide the internal structure of the blocks and how items flow between the ports. This internal structure, as shown in Figure 7.6, is consistent with the Sequence diagram in Figure 7.5.

Domain knowledge has been used in the construction of the two diagrams in Figure 7.6. For example, for the ATR–CA interface, the port at the ATR end is implemented by an antenna that transmits and receives electromagnetic waves and the port at the Compliant Aircraft end is implemented by a transponder. However, in general, such an implementation may not be known *a priori* or is part of the black-box subsystem purchased from and developed by suppliers. In this case, they are specified using general terms, like the examples shown in the diagrams.

7.3 ARCHITECTURE ANALYSIS

Before proposing the final architecture specification to the design team for the start of Design Definition, it is important to perform domain analyses at the same level to gain better understanding of the impact of architectural decisions that have been made and what the final recommendations should be. Up to this point, including the tutorial exercises, there are three scenarios being considered in the architecture. These scenarios form the basis of the architecture analysis in this part of the chapter. For any analysis to be meaningful, reasonable assumptions must be made to scope the analysis. These assumptions might be adjusted or even invalidated as the project continues, e.g., a scenario change after further meetings with the customer. As such, analysis needs to be updated accordingly.

The first scenario considered is the baseline scenario where all aircraft are identified as Compliant Aircraft. In this case, for simplicity and demonstration, an initial assumption is made that the existing ATCS facility is sufficient to deal with the air traffic flow. This establishes the baseline design specification. The system which is currently in operation is sufficient but could be improved to accommodate a higher tempo at the airport.

The second scenario considered is where an aircraft is detected but identified as non-compliant and is not reachable by communication with the ATC Controller. This scenario can be further classified into two sub-scenarios where the first sub-scenario is when the NCA does not wish to land, but only flies through the ATC airspace. This scenario is referred to as an inadvertent intrusion event. Based on a consequence analysis, this event could lead to disruption of service during the time that aircraft is within the ATC airspace. The maximum of this time, $t_{\text{intrusion}}$, can be calculated by:

$$t_{\text{intrusion}} = \frac{2R_{\text{ATC}}}{v_{\text{aircraft}}},$$

where R_{ATC} is the radius of the ATC airspace and $v_{aircraft}$ is the speed of the aircraft. With a 30 nmi radius, and assuming a low-speed light aircraft traveling at roughly 120 knots (nmi/hour), the ATCS is potentially facing the need of managing other Compliant Aircraft for a period up to 30 minutes to avoid air collisions. To deal with this situation, the ATMS might only need to provide minimal strategies to the ATCC for managing the Compliant Aircraft because the behaviour of the intruding aircraft is predictable (see Exercise 7.1.3.1). As a possible solution to Exercise 7.1.3.1, the ATMS might advise the ATC Controller to adjust the course of one or more aircraft that have higher collision risks. This likely would result in the recommendation of a new landing sequence for all aircraft by ATMS.

The second sub-scenario is when the NCA does not intend to leave the airspace, e.g., intends to land because of a declared emergency. This is referred to as a deliberate intrusion event. There is a variation of this scenario where the aircraft is intentionally disruptive, e.g., a hostile act that is intended to cause a denial of the ATC service to all aircraft. This is beyond the scope of the offering from the customer as written in the narrative but could provide the developer a long-term opportunity to seek future sponsored research and development for solutions to this highly stressing scenario.

In the second sub-scenario, the distance to be travelled by the intruding aircraft depends on its intention and thus, can be difficult to predict. The duration of time for which other Compliant Aircraft need to be managed might be long. Moreover, this second sub-scenario has a considerably higher safety risk because collisions will be more likely to happen with other aircraft perhaps already in the descending and landing phase. The current architecture supports the identification and continuous tracking of a deliberately non-compliant aircraft but does not show details of how ATMS provides solutions in the management of other aircraft based on the ATR track data. This will require a stakeholder meeting to decide whether the architecture, especially for the ATMS, must be recursively decomposed and iteratively refined further to provide solutions to the ATC Controller. An alternative decision might be to purchase high-end ATMS solutions with advanced aircraft management strategies. The decision-making process will require a trade-off study based on information such as the likelihood of the occurrence of these sub-scenarios and the available budget. These have not yet been discussed. Despite the above, an informed decision recommendation can still be made on ATR. Specifically, it needs the ability to track NCA in the event of deliberate intrusion and also inadvertent intrusion. This motivates a further domain analysis on radar performance, which will be detailed in Section 7.3.1.

The last scenario considered is a Compliant Aircraft with a faulty transponder. Although the baseline design does not have a means to detect such a fault from the ATCS perspective, to distinguish the scenario from the previous one, it is assumed that communication between the ATC Controller and the aircraft can still be established after the initial identification of the aircraft as non-compliant; thereby changing it to a Compliant Aircraft but perhaps required to operate under different flight control procedures than the ATCC uses for the other Compliant Aircraft. In fact, this scenario should also be further classified into two sub-scenarios. The first sub-scenario is where the Compliant Aircraft is in the flight plan. As such, this may not add additional effort in the management of other Compliant Aircrafts, but only will require continuous communication between the problematic aircraft and the ATC Controller. The second sub-scenario is more complicated. This is where an aircraft is compliant and intends to land but is not in the flight plan. This will affect the current air traffic flow as to whether the aircraft can land safely. This could entail a new landing sequence, and a rectification of how much disruption of service might be brought to subsequent flight planning. To answer these questions, a domain analysis on air traffic flow is evidently needed. This will be detailed in Section 7.3.2.

7.3.1 RADAR PERFORMANCE ANALYSIS

The current radar used in the ATC service has been and is operating successfully for operations with Compliant Aircraft. Although it is aging, the antenna assembly is intact and might only need a modest refurbishment. Complete replacement of the entire radar could be expensive in terms of both purchase and instalment into the existing ATC facilities. In the proposal from the developer, the minimum level of upgrade needed is a new receiver to properly support the prosed role of the ATMS in tracking aircraft in the ATC airspace. This also provides an opportunity to explore options to make the proposed ATR solution more capable for tracking NCA. The aged receiver for the current radar only supports a radar sensitivity of about 10^{-7} mW. Currently available affordable receivers can offer substantial improvement that would increase the radar sensitivity to 10^{-10} mW. It is appropriate then to investigate the benefits of this improvement as part of a first-level quantitative analysis of the radar for the NCA scenario.

The performance of an ATR at the black-box level, i.e., without decomposing it into the details of radar functionalities and components, can be characterised by three key attributes: transmission power, effective aperture, and radar sensitivity. The radar domain analysis is then centred on the radar maximum detection range equation, which is well-known and given as:

$$R_{max} = \sqrt[4]{\frac{P_t A_e G \sigma}{P_r (4\pi)^2}},$$

where P_t is the transmission power of the radar, A_e is the effective radar aperture that is further calculated by $A_e = (\lambda^2/4\pi) G$ where G is the gain of the antenna, P_r is the radar sensitivity, and σ is the radar cross-section of the aircraft being detected. For the baseline design where the ATR must have acceptable performance for detection of aircraft within the ATC airspace, R_{max} is then assumed to be 30 nmi. The current radar has a transmitter power of 100W (watt), an aperture of 1 m^2 (square meter), and a radar sensitivity of 10^{-7} mW (milliwatts).

In order to support an advanced ATMS as previously discussed, for timely management of the Compliant Aircraft in an event of NCA intrusion, the ATR must be able to detect NCA at a range of at least 30 nmi. Therefore, to be able to deal with intrusion events, the current ATR requires an upgrade to improve its detection range and be able to declare an aircraft as non-compliant when it enters the ATC airspace without being identified. This can be achieved for example, by a radar with higher transmission power, larger aperture, a greater radar sensitivity; or a combination of any of the above. Therefore, a trade-off study is necessary for the ATR Design Definition to decide what is the reasonable combination of improvements for the radar attributes that can be accommodated within a limited budget. The radar equation will be used for the trade-off analysis. This is a subject of future meetings with the customer for investigating an agreed-upon radar design evolution strategy.

To give an illustration of the types of analysis that must be done for these meetings, consider the following. Due to the fourth root in the radar range equation (which results from free-space path loss in a round trip for the transmitted wave), to double the detection range of the radar would require the transmission power to be increased by a factor of 16, if the aperture and sensitivity were kept fixed. Based on an initial search of available radar technologies, and without going into the detailed design, it is straightforward to derive a reasonable set of design constraints on these attributes. For the purpose of illustration, the design constraints specified below are set for the trade-off analysis and are reflected in the Block Definition diagram in Figure 7.6 as the design space for radar improvements:

- $100\,\mathrm{W} \leq P_t \leq 20000\,\mathrm{W}$
- $1\,\mathrm{m}^2 \leq A_e \leq 10\,\mathrm{m}^2$
- $10^{-10}\,\mathrm{mW} \leq P_r \leq 10^{-11}\,\mathrm{mW}$

As previously noted, improvements to receiver sensitivity are the recommended first choice for ATR design evolution. The second choice is to increase the power of the transmitter. The last choice would be to replace the antenna. The rationale for the ranking is based on the cost of procurement and installation.

7.3.2 AIR TRAFFIC FLOW ANALYSIS

The air traffic flow can be modelled by a simple linear model where all Compliant Aircraft approaching to the ATC airspace are assumed to have a uniform maximum speed, $v_{\max}$, of 240 nmi per hour (i.e., 4 nmi per minute). The critical spacing, d_c, between adjacent aircraft can then be calculated by:

$$d_c = \frac{v_{\max}}{n},$$

where n is the number of aircraft to be landed per every minute. With a requirement of one aircraft landed every 2 minutes as the peak tempo, i.e., $n = 0.5$, the critical spacing would then be calculated as 8 nmi, which satisfies a minimal safety spacing, d_s of 5 nmi, if all aircraft are compliant.

In the event of a Compliant Aircraft that is seeking to land but is not on the flight plan, the decision as to whether it is safe to allow the aircraft to enter the airspace and land can be based on the current traffic density, i.e., the number of aircraft within a unit space, 1 nmi. The traffic density, k, can be calculated by the inverse of aircraft spacing as:

$$k = \frac{1}{d}.$$

For a critical density of $k_c = 1/d_c = 0.125$ aircraft/nmi and a maximal safety density of $k_s = 1/d_s = 0.2$ aircraft/nmi, the following conditions should be analysed for the admittance of the unplanned aircraft into the ATC airspace:

1. If the new traffic density, k_n is such that $k_n < k_c$, then the peak tempo is not yet reached, it is safe to admit the aircraft and safe to land the aircraft by assigning it at the end of the landing queue;
2. If the new traffic density, k_n is such that $k_c < k_n < k_s$, then the aircraft can be admitted into the airspace, but cannot be landed until the density is reduced to below the critical density (assume peak tempo is the maximum landing rate the ground facilitates can accommodate safely);
3. If the new traffic density, k_n is such that $k_n > k_s$, then the aircraft cannot be admitted into the airspace as this will cause potential air collision.

Therefore, as part of the ATMS design, the above conditions should be part of the software specification. Further details on how the ATMS responds under each condition can be analysed as required. For example, as critical spacing is determined by maximal aircraft speed, under condition 2, it is possible to request other aircraft to slow down to a point where minimal safety spacing is still respected with a margin. The condition then transitions into Condition 1, where aircraft can land safely without violating the peak tempo.

7.4 SPECIFICATION OF THE IMPLEMENTATION MODEL

At this stage of the engineering project to deliver an ATCS to the customer and stakeholders, agreement has been reached on the essential architecture of the proposed system solution. The concept is to deliver a cost-effective solution with low impact on the current radar being used by the ATC service. The architect and engineering team have developed a technical package that is suitable for a review meeting with the technical team of the customer. The purpose of the meeting is to seek agreement on the design and implementation approach for the system proposed by the developer. This is expected to involve an iteration of the system and architecture definition processes that will refine the technical details of the solution in a way that the customer team will be in agreement with the developer. What follows next is a high-level specification for the software, hardware, and interfaces of the proposed system. The technical package and associated analyses have been the subject of the previous parts of this chapter. The system components (referred to as elements) and their interfaces are defined as configuration items using UML and SysML graphical models. This will be appropriate for formal management of the system design.

7.4.1 SOFTWARE SPECIFICATION

The ATCS software will be built and delivered by the development team. The software specification is part of the final essential system architecture. There are four operations defined for the ATR software component and three for the ATMS (not counting additional ones that the student may have introduced through the exercises). The expected interoperations have been specified in UML Sequence diagrams. Together with the Class and Activity diagrams that have been specified, these provide the software development team with a software architecture from which to write computer code. In addition, the architecture analysis also recommended further software specification for the ATMS in terms of the three scenarios that it shall be able to handle. Details are specified by the models in the technical package.

7.4.2 HARDWARE SPECIFICATION

The developer will purchase and integrate components for upgrading and refreshing the current radar being used by the customer. For cost-effectiveness and reduced impact on the customer budget, an evolutionary strategy will be followed. The current 1 m^2 antenna and 100 W transmitter will be retained. The baseline upgrade to the ATR will be a technology refresh of the receiver to support 10^{-10} mW radar sensitivity. This is a significant improvement over the current 10^{-7} mW sensitivity that will almost double the current detection range of the radar. This will be important for dealing with the NCA scenario.

The hardware architecture is primarily specified in a SysML Block Definition and Internal Block diagrams pair as shown in Figure 7.6. This details the need for an Air Traffic Radar hardware system and Air Traffic Management hardware to accommodate the associated software components as in the software specification. In addition, design constraints have been imposed on the hardware as derived from domain analysis. Specifically, a set of three design constraints on Air Traffic Radar are specified in the Block Definition diagram in the corresponding block.

7.4.3 INTERFACE SPECIFICATION

In its role as system integrator, the developer has defined and will implement the requisite interfaces for the ATCS. The basic interface specification for the ATCS has been modelled using SysML Block Definition diagrams. Two interfaces are defined for the ATR: the direct interface with the aircraft has identified the radar antenna the transponder on the aircraft. Decoding of transmitted waveforms is done by the receiver. Two interfaces have been defined for the ATMS. Behaviours of the four interfaces are specified in terms of exchange items (information, energy, or material) expected to flow between the connecting ports. Formal management of these items will be ensured through their specifications using SysML Internal Block diagrams.

7.4.4 MODEL SPECIFICATIONS FOR DESIGN DEFINITION

A SysML Requirement diagram summarising the specifications is offered in Figure 7.7. This is used for central management and to establish design requirements traceability

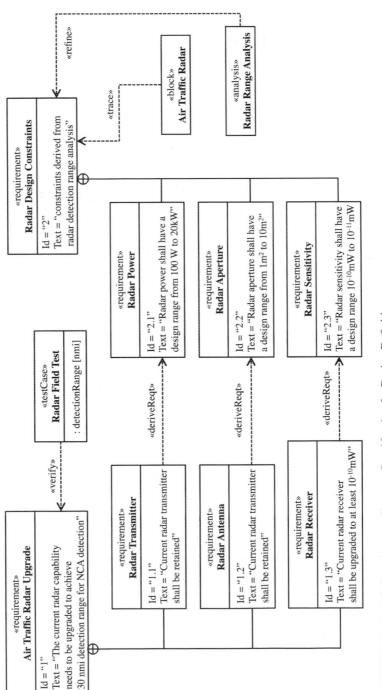

FIGURE 7.7 SysML Requirement diagram: Specification for Design Definition

to the system architecture. For the purpose of illustration, only a partial set of the specifications is summarised into this diagram.

Architecture Definition, at this point, stops in terms of further specification. However, the process is not quite complete as architecture assurance also needs to be considered for early-stage Verification and Validation (V&V), as in Figure 4.1. The benefit at this stage is self-evident: avoiding propagation of errors into the design which would be costly to rectify. However, architecture is generally difficult to verify and validate for reasons such as the reliable simulation of behaviour. As such, often only limited assurance can be achieved through methods such as expert review and model checking. The intention of this part of the tutorial is not to dive into too much detail in the V&V process, which itself is considered a large subject, but to suggest something that the architect should do in collaboration with domain engineers: preliminary test case definition. In simple terms, this process defines ways and methods in which the specifications can be tested. Following a model-based approach, these tests can be specified in UML or SysML behavioural diagrams, e.g., Activity diagrams and State Machine diagrams, with traceability to the requirements that they would verify.

7.5 SUMMARY OF MODEL SPECIFICATIONS AND TRANSFORMATION

As with the other two tutorials, this chapter summary is comprised of two parts. The first is a refinement of the system concept provided in the stakeholder narrative that is briefly summarised from the viewpoint of the system developer. The second is a specification of the types of models used in the chapter and their transformation.

7.5.1 REFINEMENT OF THE SYSTEM CONCEPT

The developer now has a detailed model of the ATCS that is ready for implementation: the ATMS and ATR elements form the first-level hierarchical decomposition of the system defined by architecture. The current radar used by the existing ATC service works well with Compliant Aircraft in the airspace. This is the baseline scenario for the ATCS solution. The alternative scenarios and architecture analysis have shown that significant upgrades to the current radar will be needed for the more stressing scenarios. The substantial investment required cannot be justified at this point in the upgrade of the ATC service. Retaining the existing 1 m^2 antenna and 100 w transmitter but upgrading the receiver as part of the ATR integration with the proposed ATMS is the recommended radar solution. An engineering analysis will be performed to determine if the antenna needs to be refurbished (not replaced) and if the transmitter needs a technology refresh. ATR design evolution will include a transmitter upgrade as required by and agreed upon with the customer. An antenna upgrade can be considered in the long-term evolution but this should be part of a strategic plan for the ATR that the developer seeks to define in collaboration with the customer.

The design driver for the proposed ATCS solution is the ATMS. In the baseline scenario, the reduced workload offered to the ATCC can allow the airport to increase its tempo and thereby usage of the airport. The ATMS will be configured with a cost-effective but powerful computing system from the beginning of its implementation. The software design will be evolved progressively to reduce the fiscal impact on the airport. The disruption of (ATC) service (DoS) scenarios will be addressed through the development of algorithms to assist the ATCC in rapid redirection of the flight patterns of participating Compliant Aircraft in the ATC airspace. In the first scenario, the algorithms can be invoked whenever a fleeting NCA is detected. Better integration with the regional FPMS can improve situational awareness and possibly alert the ATCC to possible intrusions. The DoS in this scenario can be as long as 30 minutes. This will be the first stage of design evolution of the ATMS.

The second stage of design evolution will be further enhancement of the algorithms and decision logic to accommodate the management of the ATC airspace in the event of a Compliant Aircraft entering the controlled airspace with a faulty transponder. This will assist the ATCC with the challenges of making a rapid decision as to whether to permit the aircraft to declare an emergency and continue its approach then land at the airport. The disruption in this second DoS scenario is expected to be in the range of 10–15 minutes. Decision logic will include but not be limited to considerations of current tempo at the airport as well as local weather and visibility. Again, integration of the ATMS with the regional FPMS can facilitate dealing with this scenario through better understanding of rerouting options for the problem aircraft.

A long-term research and development effort will also be conducted to work with the customer on the total denial of service scenario when an intentional intrusion into the ATC airspace is made by an air vehicle.

The state-of-the-art qualified model-based systems engineering process used by the developer will ensure the integrity of the ATCS design and its alignment with the voice of the customer as it evolves through the proposed stages of evolution.

7.5.2 MODEL TYPES AND TRANSFORMATION

The tutorial in this chapter complements the implementation of the Framework for Structured Analysis and Design in the previous two tutorials (Chapters 5 and 6) by introducing a process to extend the baseline behaviour of the system for alternative scenarios then integrate alternative system behaviours. The process of model specification and transformation employed in this last tutorial is summarised using the results presented in Figures 7.1–7.3 and Table 7.1. The aim of the process is to refine the essential architecture and system specification to a level of detail that is suitable for design and, if sufficient, for implementation. The following summary highlights the progression of model specifications and transformation employed in the process.

The first stage in the progression is an iteration of the functional decomposition process. Previously this was concerned with decomposing a use case into included use cases. Here, the focus shifts to decomposing a use case into extending use cases, as depicted in Figure 7.8. Each extending use case is considered to be a next-level function of the extended use case because its execution relies on the extended use

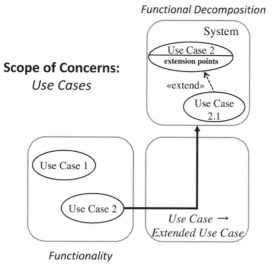

FIGURE 7.8   Iterative functional decomposition for alternative behaviour

case but will only occur when specified conditions are met at the extension point. In this way, the different behaviours of the system and subsystems are automatically integrated at the level of functionality, as captured in a set of integrated system and subsystem level Use Case diagrams, e.g., as in Figure 7.1.

The next stage is concerned with transforming the Use Case diagrams into an Activity diagram in which the alternative behaviours are modelled as alternate flows additive to the basic flow. The integrated model in Figure 7.9 exhibits an exclusive

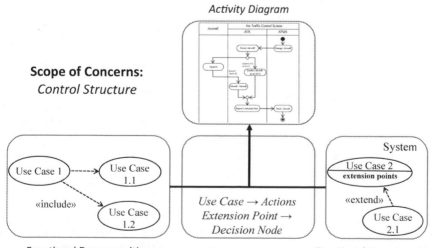

FIGURE 7.9   Specification of integrated behaviours

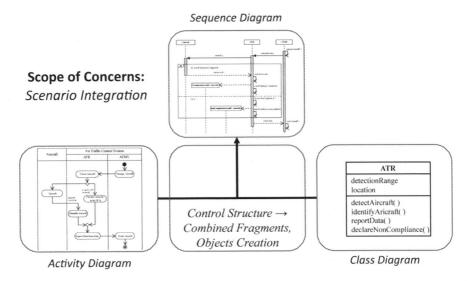

FIGURE 7.10   Synthesis of essential architecture with alternative behaviours

choice control structure. For real-world complex systems, the control structure complexity can grow quickly as the number of alternative behaviours increases and can become nested with inner feedback loops and concurrent flows. Complexity management is a strong motivation for the explicit modelling of alternative behaviours through iteration rather than considering multiple behaviours concurrently starting from System Definition. It is worth noting that the Use Case description is also applied prior to this stage, but this detail is suppressed here.

The final stage is concerned with synthesising the integrated Activity diagram into the Sequence diagram with revisions made to the Class diagram. This is depicted in Figure 7.10. In the ATCS tutorial, this evolves the control structure into combined fragments with "if...then; else" statements and with object creation. In other situations, depending on the detailed set of behaviours, the Sequence diagram will have other types of structure.

With a sufficient level of detail in the Essential Architecture, interface identification and definition can be carried out as demonstrated in Section 7.2 of this chapter. Subsequently, with a sufficient level of architecture and domain analysis, the specification can then be made and agreed upon to enable detailed design, which is the next stage in the Systems Engineering lifecycle depicted in Figure 4.1 in Chapter 4.

7.6 THE ART OF SYSTEM ARCHITECTING AND DESIGN

Conceptual integrity is the most important consideration in system design. The architect should therefore play a central role throughout the design and implementation of a system. System architecting and design are more than just a subject of science and engineering. They are a cognitive activity for problem-solving in which knowledge is transformed into solutions. This is an art. As summarised by Wilkinson and noted in the brief history of architecture in Chapter 1, Section 1.1, architecture is a conception in the imagination and indeed for a system design is the highest-level conception. It is inherently subjective.

Architectural methods for product and service development will drive the engineering solution but not in a unique way. This poses a challenge to practicing architects and engineers that can also be observed in the teaching of system architecture. For the same problem, even with the same starting point, architectural solutions proposed by students can differ greatly, especially as the details of the system design evolve. Although two solutions can both be technically correct, one *representation* of a system can be considered better than another if it has certain desired qualities. One key quality is simplicity: minimal detail in an architecture description of a system without reducing the information content to less than what is essential. If architecture is to be useful it must be simple.

Model complexity grows with system complexity, but a complex graphical model defeats the intended benefit of using a model-based approach to facilitate stakeholder communication. Diagrams should be simple and intuitive. In the tutorials, models have been specified in the simplest way that is practical by using structured design to reduce complexity. For student readers, model complexity will be observed in the first practical case study (Chapter 11). The second practical case study (Chapter 12) provides readers and students an opportunity to explore the art of system architecting by utilising the modelling concept called composite structure, as explained in Chapter 10, to reduce complexity. This is just one such architectural method amongst many. The three tutorials and the two practical case studies have been written with the art of system architecting and design in mind with the intention of helping readers and students understand how this art should be integral to the practice of modern engineering.

Part III

Modelling Languages

OVERVIEW OF THE MODELLING LANGUAGES

The purpose of the three chapters in this part is to introduce two of the most widely used modelling languages in the model-based systems engineering community. These are the Object Management Group's Unified Modeling Language (UML) and Systems Modeling Language (SysML). This part is primarily intended for the general reader or student who has little experience with and knowledge of these modelling languages. As such, instead of covering the whole spectrum of diagram types, the scope is a core set of diagrams that facilitate the model-based approach introduced in Part I and demonstrated in the tutorial case study in Part II.

Each chapter in this part is organised by diagram type. Chapters 8 and 9 are focused on UML and Chapter 10 on SysML. Within each part of each chapter, a specific UML or SysML diagram type is introduced in terms of the core model elements required along with explanations of the syntax, semantic and modelling rules. For the learning of UML, prescriptive modelling procedures are offered using a running example for the modelling of a simple automated teller machine that illustrates, for each type of diagram, how the model elements should be specified and structured. A running exercise on the modelling of an Online Shopping System is further offered to complement the modelling procedures prescribed. For the learning of SysML, a scoped example of vehicle subsystems modelling is provided with exercises on improving and completing the system model. In addition to the practical exercises, toward the end of each chapter, further exercises are offered to challenge the student to understand the linkage between architecting and system modelling. This includes gaining insight as to how desired qualities, such as model consistency, can be achieved. The student is, therefore, led in bite-sized steps to develop practical skills that can be used for architecting and system development, not just for the drawing of diagrams.

8 Modelling Languages I
Functionality and Behaviour

KEY CONCEPTS

Graphical models
Modelling languages
Modelling of functionality
Modelling of functional flows

The aim of the first two Modelling Languages Chapters is to introduce the Unified Modeling Language™ (UML)[1]. As compared to the published standards of UML, these two chapters offer a distilled version of how to use its modelling concepts and elements to model and represent a system with UML diagrams. The concepts and knowledge covered in this chapter and the next provide the necessary toolkit for accomplishing the essential activities of System and Architecture Definition.

It is a widely adopted practice in the teaching of UML to start with its core modelling concepts, Class and Classifier, and hence the Class Diagram. However, for our voyage into systems engineering, the development of a system will begin with its required functionalities. The journey into UML in this book, therefore, starts with modelling functionality and behaviour. This aligns with the first Case Study Tutorial as presented in Chapter 5.

The words *model* and *diagram*, particularly in the context of modelling languages, are often inappropriately used, as if they are interchangeable concepts. Model, as discussed in Chapter 2, can be defined as an abstraction that is an interpretation of a system concept; diagram, on the other hand, is a representation of a specific perspective of a model. Therefore, it is appropriate to say that a model of a system can be represented by a set of diagrams; in particular, a UML class can represent an element of the system (Dickerson 2013). In model-based approaches, it has been a common practice that the modelling of systems is centred around constructing diagrams where the creation of a graphical element in a diagram also specifies the corresponding model element in the model. The model of a system is then completed by building diagrams that capture different perspectives of the system. As such, UML diagrams constructed in this way are often called *graphical models*.

To help the readers and the students to distinguish these two interrelated concepts; in this book, Pascal Case, i.e., concatenating capitalised words, is adopted for calling the names of a UML model element, e.g., UseCase, and in italics when first introduced; normal (lower) cases for calling and referring to a specified graphical element, e.g., the "Make Payment" use case; and capitalised cases for calling a UML diagram, e.g., the "ATM Functionalities" Use Case diagram. For simplicity, once a

DOI: 10.1201/9781003213635-8

graphical element has been specified, the name of the use case, e.g., "Make Payment" will also be used to refer to this use case within the same modelling context.

The practice of modelling languages for systems architecting presented in this chapter is organised as follows:

A Brief Overview of UML
Use Case Diagram
Activity Diagram

8.1   A BRIEF OVERVIEW OF UML

The Unified Modeling Language™ (OMG 2017b), standardised and published by the Object Management Group (OMG), is the *de facto* standard for object-oriented software development. It is a graphical language that provides modelling syntax (graphical notations) and semantics (meaning of notations) for object-oriented problem solving. Models are important to software development for both engineering and communication, just like how blueprints drawn by architects are used in the construction of buildings. The more complicated the building, the more critical the communication between the architect and builder, and the architect and the customer. One of the challenges in such critical communications is to ensure the intention of what is communicated can be interpreted correctly. This is what motivated the effort for unifying modelling languages in the domain of software engineering to ensure that the software engineers 'speak' the same and referenceable language. Eventually, this led to the birth of UML. There have also been other objectives set by the UML community as the modelling language evolved, for example, the idea of using UML to create machine-readable software architecture that can be transformed into computer code for rapid deployment.

Since UML was originally developed for software engineering, why is it introduced and used here for systems engineering? Simply put, it fits the purpose of model-based systems engineering. The fundamental and core concepts of UML enable engineers to model system elements and interactions between the elements, whether it is a software system, an electrical system, a mechanical system, or so on and so forth. Arguably, UML might be insufficient for modelling systems in a specific domain other than the software one. This, however, is compensated by its extensibility and customisability through the so-called *stereotype* (denoted by «stereotype name», e.g., «include»), which enables the creation of domain-specific notations and terminologies extended from the core elements of UML. The collection of a set of stereotypes for the purpose of specific domain modelling or problem solving is referred to as a UML *profile*. Indeed, due to the primary focus on software engineering, a customised version of UML for general systems engineering was developed beginning in the early 2000s and standardised by the OMG. This is known as the System Modeling Language™ (SysML) (OMG 2017a). We will come back to SysML at the end of this chapter as to why UML is preferred as the starting point for architecting in this book.

As discussed, UML defines the syntax and semantics of modelling elements. It then organises these elements into two types of diagrams, structure diagrams and behaviour diagrams, to provide means of representing the structure and behaviour of the subject (system), respectively. Of all UML diagram types, this book will cover the Use Case, Activity, and Sequence diagrams for behavioural representation and the Class diagram for structural representation. These are the essential diagrams for completing architecture and system specifications. Nonetheless, this is not to say that the rest of the diagrams and their associated modelling elements are not relevant in the overall systems engineering processes.

There are over a dozen tool vendors offering the capability of creating UML and SysML models and diagrams ranging from commercial level ones such as Artisan Studio, Enterprise Architect, MagicDraw, and Rational Rhapsody to open sources ones such as Eclipse. To complete the exercises and the case studies in this book, it is essential that the student chooses a preferred modelling environment to work with. In principle, there is no reason why one cannot create UML diagrams in software such as PowerPoint. However, there are many advantages for using a professional tool. For example, it makes it easier to organise model elements for traceability and reusability.

Exercise 8.1.1: UML Project

1. Select a tool of personal/organisational preference and familiarise your-self with the user interface.
2. Using this tool, create a new UML project with the name "Learning UML".
3. Discover how to create UML diagrams within this project.

8.2   USE CASE DIAGRAM

The Use Case diagram is the most widely used of all UML diagrams. Although one of the simplest offered by UML, it provides a powerful modelling capability. It intends to show the interactions between a system of interest and the users or other external systems in the environment. The viewpoint of a Use Case diagram should be that of an external observer, which leads to a 'black-box' view of the system. The emphasis is on *what* the system does rather than *how* it does it. As such, it became a popular method for requirements definition and an executive level means to facilitate the communication between stakeholders and system architect.

This part of the chapter explains how to create a Use Case diagram to capture the system boundary, elements in the system environment that interact with the system, and system functionalities. A list of graphical elements used in the construction of a Use Case diagram to achieve the above is depicted on the left in Figure 8.1. An exemplar "ATM Functionalities" Use Case diagram is depicted on the right in the figure, where ATM is the acronym for automated teller machine.

Exercise 8.2.1: UML Use Case Diagram

1. Under the UML project created in Exercise 8.1.1, create a new Use Case diagram.
2. Discover how to give this diagram a name and where the name is displayed.
3. Name the Use Case diagram as "Learning Use Cases".

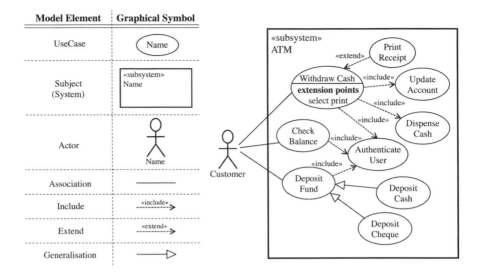

FIGURE 8.1   UML Use Case diagram notations and example: ATM Functionalities

A well-constructed Use Case diagram should contain the following model elements: The system, which is referred to as the *Subject* in UML; a set of *UseCases* that capture system functionalities; and a set of *Actors* representing elements in the system environment that interact directly with the system as specified by the use cases. Let us now look at the graphical notations and common usage of these three model elements individually.

In a Use Case diagram, the system is bounded by a rectangle box, with its name displayed in the top-left corner,[2] e.g., "ATM". Simple as this may sound, this is very important as this box strictly defines the boundary and scope of the system of interest; and everything depicted outside of this box is considered as external to the system. Therefore, one can assume that a system designer has no design authority on everything that is outside of this box; and the design of the system is constrained by the presence of these external elements given that they are interacting with the system.

The system environment can be very large and complex. However, as discussed in Chapter 2, a good model should only consider elements in the system environment that are interacting with the system. In Use Case diagrams, these are modelled by Actors and often depicted as stick figures. This does not necessarily mean that an Actor must be a human role, e.g., a customer or maintenance engineer; it can also be other physical elements (systems) such as a vehicle or an airplane; or an even abstract element, e.g., a regulation or even a weather condition. A good naming convention of an Actor should not include an article and should always be singular.

Inside the system box, system functionalities are modelled by UseCases. A UseCase has an oval shape containing a short description of the specific usage (system functionality). This description is often referred to as the "name" of the UseCase. Although UML does not provide specific rules on how to name a UseCase, a general rule of thumb is that a UseCase name should be constructed based on a verb-noun phrase using first person present tense and singular noun, e.g., "Withdraw Cash".

Exercise 8.2.2: Elements in a Use Case Diagram

Within the "Learning Use Cases" Use Case diagram created earlier,

1. Create a subject (system boundary) and name it as "System".
2. Create an actor with the name, "Actor". This actor should be outside of the system boundary.
3. Create a use case with the description, "Use Case". This use case should be inside of the system boundary.
4. If a professional tool is used, while creating graphical elements in the Use Case diagram, observe how model elements are created and stored in the project. Play around with creating/deleting model elements as compared with creating/deleting graphical elements.

A Use Case diagram containing only the elements discussed above is hardly meaningful. This is because it has not yet captured any existing *Relationships* between the elements. The semantics of UML Actor and UseCase explained so far suggest the need of an interaction between a specified actor and the system to describe the fact that the actor 'uses' the system in the way prescribed in the specified use case. For example, the "Customer" actor 'uses' the ATM "Check Balance" use case (functionality), as depicted in Figure 8.1. In UML, this interaction is modelled by an *Association* relationship. Every use case specified should be associated with at least one specified actor. This association can be indirect, and this will be made clearer later.

Exercise 8.2.3: Association

Continue with the Use Case diagram drawn so far,

1. Connect the actor and the use case created in Exercise 8.2.2 using the Association relationship.
2. If a professional tool is used, observe how the model element "Association" is specified and stored in the project. Check its properties.

Having understood the basic structure of a Use Case diagram, create a new Use Case diagram under the "Learning UML" project, named as "Online Shopping System". In this diagram:

3. Create an "Online Shopping System" system with a "Customer" actor.
4. Create a "View Shopping Cart" use case.
5. Connect the "Customer" actor and the "View Shopping Cart" use case with an association to show the interaction in which the customer uses the "View Shopping Cart" functionality of the Online Shopping System (OSS).

There are three additional types of UML Relationships available for use in constructing a Use Case diagram that will be introduced and explained next.

Include, with a direction going from the including UseCase to the included UseCase, is used to model two situations. The first situation is to simplify a complex (including) UseCase into several (included) lower-level UseCases in which the lower-level UseCases are necessary for a complete description of the complex UseCase. For example, as illustrated in the ATM Use Case diagram in Figure 8.1, the complete description of "Withdraw Cash" involves all of the three lower-level use cases, "Authenticate User", "Update Account" and "Dispense Cash". The second situation is to extract the commonality between multiple (including) UseCases into a separate (included) UseCase. For example, as illustrated in the same Use Case diagram, "Authenticate User" is a use case common among the three higher-level use cases, "Withdraw Cash", "Check Balance" and "Deposit Fund", and is therefore modelled as an included lower-level use case. In both situations, included UseCases is considered as the next-level decomposition of the including UseCase and must be executed for a complete execution of the including UseCase. Therefore, included UseCases inherit the Association relationship between the including UseCase and its associated Actors. As such, to manage model complexity, it is unnecessary to create an explicit Association between the involved Actors and the included UseCase, unless the included UseCase interacts with the Actor(s) in a different manner. This is an example of an indirect relationship, as mentioned earlier.

Exercise 8.2.4: Include

In the "Learning Use Cases" Use Case diagram,

1. Create an "Including Use Case" use case.
2. Create an "Included Use Case" use case.
3. Connect the two use cases by an «include» relationship, in which the direction of the arrow is pointing toward the "Included Use Case" use case.

Having understood how «include» relationship is drawn, in the "Online Shopping System" Use Case diagram,

4. Create a "Make Payment" use case that specifies the system interaction with the "Customer" actor.
5. Create a "Confirm Payment" use case and a "Generate Order" use case, both *included* by the "Make Payment" use case.
6. Create an "Add Item to Shopping Cart" use case and a "Remove Item from Shopping Cart" use case; both interact with the "Customer" actor.
7. Create an "Update Shopping Cart" use case that is *included* by both of the use cases created in Step 3 above.

Extend, with a direction going from the extending UseCase to the extended UseCase, is used to model additional, often optional, behaviour to an existing (extended) UseCase under a specific condition. This specific condition is modelled as a unique *ExtensionPoint* under the extended UseCase. As depicted in

Figure 8.1, "Print Receipt" as an extending use case, is optional and will only be performed when the customer selects to print the receipt. In this case, the specific condition is modelled by the "select print" extension point. From this example, it is evident that the extended UseCase is meaningful independently of the extending UseCase while the reverse is not true: the customer cannot print a receipt without withdrawing cash. The extending UseCase inherits the Association relationships owned by the extended UseCase. Further, because an extending UseCase is an additional functionality required to complete the optional usage of the system, additional Actor(s) may be involved in the execution of the extending UseCase.

Exercise 8.2.5: Extend

In the "Learning Use Cases" Use Case diagram,

1. Create an "Extended Use Case" use case.
2. Create an "Extending Use Case" use case.
3. Connect the two use cases by an «extend» relationship, in which the direction of the arrow is pointing toward the "Extended Use Case" use case.

Having understood how «extend» relationship is drawn, within the "Online Shopping System" Use Case diagram created earlier,

4. Connect the "Remove Item from Shopping Cart" use case and the "View Shopping Cart" use cases by an «extend» relationship; which direction should the arrow be pointing if the "View Shopping Cart" use case is *extended* by the "Remove Item from Shopping Cart" use case?
5. Modify the "View Shopping Cart" use case to include the corresponding extension point, "unwanted item".

Generalisation, within the context of Use Case diagrams, is used to model a situation where a set of specific use cases can be generalised into a single use case that abstracts the behavioural commonality. As shown in Figure 8.1, "Deposit Cash" and "Deposit Cheque" use cases are generalised into a more abstract "Deposit Fund" use case. This situation is modelled by using the Generalisation relationship going from the two *child* use cases to the *parent* "Deposit Fund" use case, with the hallow triangle attaching to the parent use case. Generalisation can also be used to model structural commonality of Actors. A more detailed and formal account of UML Generalisation will be provided in Chapter 9 where UML Classifier is introduced.

In practice, the usage of the include relationship is sometimes confused with generalisation. Generalisation does not, in principle, create a meaningful system hierarchy, but can create a useful model hierarchy to manage model complexity. For instance, without the "Deposit Fund" use case, both "Deposit Cash" and "Deposit Cheque" use cases will need to include the "Authenticate User" use case, making the diagram unnecessarily more complicated.

Exercise 8.2.6: Generalisation

In the "Learning Use Cases" Use Case diagram,

1. Create a "Parent Use Case" use case.
2. Create a "Child Use Case" use case.
3. Connect the two use cases by a Generalisation relationship, in which the direction of the arrow is pointing toward the "Parent Use Case" use case.

Having understood how the Generalisation relationship is drawn, within the "Online Shopping System" Use Case diagram created earlier,

4. Create a "Modify Shopping Cart" use case.
5. Connect this use case to the previously created "Add Item to Shopping Cart" and "Remove Item from Shopping Cat" use cases to show that the "Modify Shopping Cart" use case *generalises* the other two use cases. Which direction should the two arrows be pointing? Also note how the "Modify Shopping Cart" use case creates a model hierarchy but not a functional hierarchy.
6. With the introduction of the "Modify Shopping Cart" use case, the two associations that previously connected the two child use cases to the "Customer" actor now need to be removed and replaced by associating the "Customer" actor with the "Modify Shopping Cart" use case. The two child use cases then inherit this new association relationship such that the previously (removed) associations are 'kept' in this manner.

8.3   ACTIVITY DIAGRAM

A UML Activity diagram provides a means to present the *flows* in a system. In essence, an Activity diagram is formed by a combination of nodes (*ActivityNodes*) and directed edges (*ActivityEdges*) that connect the nodes. Depending on the meaning of the system element that the node is representing, different model elements need to be used.

A list of model elements used in the modelling of flows in an Activity diagram is depicted in Figure 8.2. It is worth noting that this is not a full list of model elements defined by the UML standard for an Activity diagram, but it covers the basics. More advanced modelling concepts, such as interruption and signals, build on top of these basic ones. An example Activity diagram is depicted in Figure 8.3 involving the usage of most of these model elements. This Activity diagram is an elaboration of the "Check Balance" use case of the ATM in Figure 8.1. (Elaboration, as seen in the tutorial case studies, is one of the underlying concepts in the transformational process of architecting.)

Exercise 8.3.1: UML Activity diagram

1. Under the "Learning UML" project created in the early exercises, create an Activity diagram.
2. Discover how to give this diagram a name and where the name is displayed.
3. Name the Activity diagram "Learning Activities".

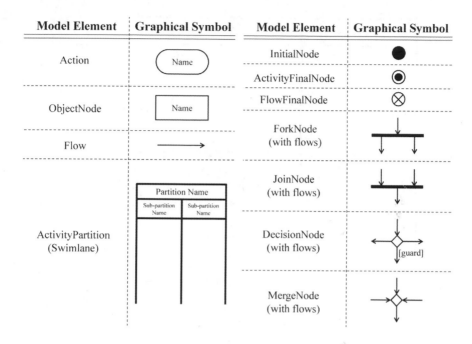

Model Element	Graphical Symbol	Model Element	Graphical Symbol
Action	Name	InitialNode	
		ActivityFinalNode	
ObjectNode	Name	FlowFinalNode	
Flow	→	ForkNode (with flows)	
		JoinNode (with flows)	
ActivityPartition (Swimlane)	Partition Name / Sub-partition Name / Sub-partition Name	DecisionNode (with flows)	[guard]
		MergeNode (with flows)	

FIGURE 8.2 UML Activity diagram notations

The fundamental building blocks of an Activity diagram are *Actions*, drawn in rounded rectangles. According to UML rules, an Action is not further decomposable within the Activity that contains it, but itself can be another Activity. This enables the representation of system behavioural hierarchy through a set of layered Activity diagrams instead of a single Activity diagram containing nested levels.

When connecting a precedent Action to its subsequent Action by a *ControlFlow* (directed arrow), a control flow is formed. This models the fact that the execution of the subsequent Action is initiated when and only when the precedent Action finishes its execution. How does the subsequent Action know when this happens? In real-world systems, this is achieved for example, through a message notification or an output-to-input relationship. As such, if what is being passed on from actions to actions is to be modelled explicitly, an *ObjectNode*, notionally drawn as a rectangle containing the description of the object must be used. An *ObjectFlow* is therefore a flow that involves the creation or consumption of objects, represented by an Action flowing into an ObjectNode, then flowing into another Action. The above discussion shows that UML categorises *Flows* into ControlFlow and ObjectFlow due to its semantic necessity, e.g., for the specification of detailed execution mechanisms in computer programming. However, in modelling general systems, the differentiation is not critically needed. The structure of the flow is what is of most importance to architecting.

In specifying different flow structures, e.g., series, parallel, alternate, and loop, various *ControlNodes* are introduced by UML. The following explains when and how to use each of these control nodes.

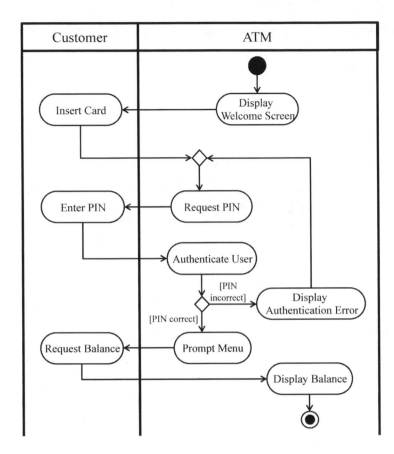

FIGURE 8.3   UML Activity diagram example: ATM Balance Checking

The beginning of a flow is always marked by an *InitialNode*. When it comes to the end of a flow, there are two possible situations. If there is only one flow ending point in the modelled Activity, like the example in Figure 8.3, an *ActivityFinalNode* is used. If there are multiple flows that have different ending points in the modelled Activity, an *ActivityFinalNode* is used when reaching here and terminates the entire Activity. Otherwise, a *FlowFinalNode* is used, which only terminates the flow that proceeds to this end point; and there might be other still on-going flows in the Activity. There is a simple rule of thumb to avoid confusion with which type of final node to use: only use the ActivityFinalNode for the intended flow (also referred to as the basic flow) and use FlowFinalNodes for all other flows.

Exercise 8.3.2: Series Flow

Within the "Learning Activities" Activity diagram created earlier,

1. Create two actions named as "Action 1" and "Action 2".

2. Connect the two actions by using a control flow (or just flow in some tools) with a direction going from "Action 1" to "Action 2".
3. Create an initial node and an activity final node.
4. Connect the two control nodes to the appropriate actions, such that the Activity starts with "Action 1" and terminates when "Action 2" is completed.

Having understood how series flow can be presented in an Activity diagram, create a new Activity diagram, named "Online Shopping". In this diagram, model the scenario for the "Add Item to Shopping Cart" use case described as follows: the Activity starts with the Customer selecting an item to view followed by the OSS providing information of the selected item. Then, following the Customer adding the item to the shopping cart, the OSS updates the shopping cart accordingly. To complete this model, perform the following steps:

5. Create a list of actions: "Select Item to View", "Provide Information", "Add Item to Shopping Cart" and "Update Shopping Cart".
6. Connect these four actions using control flows in series following the order as presented.

To model parallel flow structures, i.e., multiple flows that are happening concurrently in the same period of time, *ForkNode* and *JoinNode* are used. The two nodes use the same graphical notation, a solid bar, but with the ForkNode having only one incoming Flow and multiple outgoing Flows and the JoinNode having only one outgoing Flow and multiple incoming Flows. The multiple outgoing and incoming Flows are semantically concurrent flows. Based on this semantic, for model clarity, it is therefore advised to use them as a pair whenever possible, although this is not formally specified in the UML standard.

Exercise 8.3.3: Concurrent Flow

Within the "Learning Activities" Activity diagram created earlier,

1. Create two new actions named as "Parallel Action 1" and "Parallel Action 2" respectively.
2. Create a fork node and join node pair.
3. Revise the previous flow structure such that:
   a. "Action 1" flows into the fork node;
   b. the fork node then flows into both of "Parallel Action 1" and "Parallel Action 2";
   c. both "Parallel Action 1" and "Parallel Action 2" further flow into the join node;
   d. and finally, the join node flows into Action 2.

Having understood how concurrent flow can be presented in an Activity diagram, in the "Online Shopping" Activity diagram created earlier, model the elaborated "Make Payment" use case of the OSS to reflect the following scenario: The customer makes the payment following a payment request from the OSS. Once the payment is completed, the OSS concurrently confirms the payment and generates an order. To complete this model, perform the following steps:

4. Create the following new actions: "Request Payment", "Make Payment", "Confirm Payment" and "Generate Order".
5. Create a fork node and join node pair.
6. Connect the specified actions and the nodes following the flows given in the narrative: "Confirm Payment" and "Generate Order" actions need to connect to the pair of fork node and join node to reflect concurrency.

To model alternate flow structures, i.e., flow that happens conditionally, *DecisionNode* and *MergeNode* are used. The two nodes use the same graphical notation, a diamond, but with the DecisionNode having only one incoming Flow and multiple outgoing Flows, and the MergeNode having only one outgoing Flow and multiple incoming Flows. The multiple outgoing and incoming Flows are semantically alternate flows. The Activity flows through one of these alternate flows only when the associate condition(s) is met. Conditions are modelled by *guards*, represented as a description on the outgoing edges from the DecisionNode within a square bracket, e.g., [PIN incorrect]. As in the ForkNode and JoinNode case, using the DecisionNode and MergeNode as an accompanying pair is also recommend. However, this is not always practical because quite often, an alternate path may never merge back with the other alternate paths and may eventually flow into a FlowFinalNode.

It is also possible to model loop structure by using a DecisionNode and MergeNode pair, with the MergeNode appearing prior to the DecisionNode. An example of this is shown in the Activity diagram in Figure 8.3, where the loop for user authentication is modelled in this way. However, it is evident that this is not sufficiently detailed, e.g., the exit condition of such a loop is not specified. This naturally forces further elaboration of an Activity into more detailed behaviour model represented in a UML Sequence diagram, where the modelling of loop structure is better facilitated.

Exercise 8.3.4: Alternate Flow

Within the "Learning Activities" Activity diagram created earlier,

1. Create a new Action named as "Alternate Action".
2. Create a pair of decision node and merge node.
3. Revise the previous flow structure such that:
   a. the Activity starts with "Action 1", which flows into the decision node;
   b. the decision node then flows into both of the folk node and "Alternate Action";
   c. the previous concurrent structure between "Parallel Action 1" and "Parallel Action 2" remains unchanged;
   d. the join node, instead of flowing into "Action 2" directly, now flows into the merge node;
   e. the "Alternate Action" also flows into this merge node;

     f.   and finally, the merge node flows into "Action 2", which by its completion terminates the Activity.

4. Specify the guard as "[x=1]" for the out-going edge from the decision node to the folk node.

The final flow structure should have two alternate paths with one of the paths containing a set of two concurrent paths. Having understood how alternate flow can be presented in an Activity diagram, in the "Online Shopping" Activity diagram created earlier, model the behaviour of the "View Shopping Cart" use cases with the extension "Remove Item from Shopping Cart" to reflect the following scenario: The customer starts with viewing the shopping cart followed by the OSS presenting the shopping cart to the customer and then requests confirmation from the customer. If the customer confirms the shopping cart, he or she will proceed to checkout. If the customer does not want a particular item in the shopping cart anymore, the customer would remove the item from the cart. The OSS then responds by updating the shopping cart and presents the updated version back to the customer for confirmation. To complete this model, perform the following steps:

5. Create the following new actions: "View Shopping Cart", "Present Shopping Cart", "Request Confirmation" and "Proceed to Checkout".

6. Connect these actions using control flows in series following the order as presented. This partial flow models the basic flow when no extension ("Remove Item from Shopping Cart") presents.

7. Create a "Remove an Item" action, make a copy of the "Update Shopping Cart" action created in the previous exercise, and create a decision node and merge node pair.

8. Revise the basic flow such that "Request Confirmation" now flows into the decision node which then branches into two alternate flows: the original flow that lead to "Proceed to Check-out", and a new flow that proceed to "Remove an Item" with the guard content, "unwanted item". The new flow continues with "Update Shopping Cart", then flows into the merge node which finally flows back to "Present Shopping Cart".

9. Finally, the flow from "View Shopping Cart" to "Present Shopping Cart" also needs to be revised such that "View Shopping Cart" now goes through the created merge node before proceeding into "Present Shopping Cart". Hint: use the example Activity diagram in Figure 8.3 as a reference for the exclusive choice (decision-to-merge) control structure.

Finally, let us look at how ActivityNodes with common characteristics can be grouped by using *ActivityPartition*. An ActivityPartition is notionally represented by a so-called swim lane, as shown in Figure 8.2 with an example usage shown in Figure 8.3. In the "ATM Balance Checking" Activity diagram, actions that are performed by the customer are grouped under the "Customer" activity partition, while actions that are performed by the ATM are grouped under the activity partition, named "ATM". This way, the application of ActivityPartition to the existing ActivityNodes provides a clear definition of the viewpoint of the Activity diagram.

ActivityPartitions also facilitate system decomposition by including *subpartitions* represented by inner-swim lanes contained by the parent swim lane. For instance, imagine the decomposition of the ATM into a Console subsystem, a Display Unit subsystem and a Receipt Printer subsystem; this would lead to modelling three inner swim lanes within the "ATM" swim lane.

Exercise 8.3.5: Partition

Within the "Learning Activities" Activity diagram created earlier,

1. Create an activity partition with two swim lanes with the names, "System 1" and "System 2" respectively.
2. Without changing the flow structure, move the "Alternate Action" into the "System 1" swim lane and rest of the actions and control nodes into the "System 2" swim lane.
3. Within the "System 2" swim lane, create two inner swim lanes, named as "Sub-system 1" and "Sub-system 2" respectively.
4. Without changing the flow structure, separate the actions in the "System 2" swim lane such that "Action 1", "Action 2", the initial node, the activity final node, the decision node, the merge node and the folk node are all allocated to "Sub-system 1" while "Parallel Action 1", "Parallel Action 2" and the join node are allocated to "Sub-system 2". Note that in terms of architecting, allocation of control nodes reflects where control takes place.

Having understood how Partition works, in the "Online Shopping System" Activity diagram,

5. Create an activity partition with two swim lanes with the names, "Customer" and "OSS" respectively.
6. Based on the narratives provided in the previous three exercises, without changing the flow structure, allocate all actions created so far to the appropriate partition. All control nodes apart from the decision node should be allocated to the OSS as the decision of whether to proceed to checkout or to remove an item is made by the customer, while the different branches are merged by the OSS.
7. Connect the partial scenarios from the previous three exercises in series to form the complete flow for the "Online Shopping" Activity, i.e., "Update Shopping Cart" flows into "View Shopping Cart" and "Proceed to Checkout" to "Request Payment". No new actions are needed.
8. Finally, complete the model by adding and allocating an initial node to the "Customer" partition, and adding and allocating an activity final node to the "OSS" partition.

To summarise, it is clear from the running example of ATM and the OSS exercise that *Activities* are a means to model system behaviour. In both cases, Activity diagrams are created with reference to the Use Case diagrams. As such, the Activity diagrams took a mixed viewpoint from actors that are external to the system as well as elements internal to the system. In principle, Activity diagrams are not restricted to such a viewpoint, e.g., an Activity diagram can illustrate the behaviour of a subsystem in absence of its interaction with other subsystems, thereby showing a completely internal viewpoint of this subsystem.

It was also evident that an Activity diagram is insufficient to present the complete, detailed behaviour of a system. Therefore, other diagrams and model elements are necessary to capture other essential system elements and relationships. In addition to the loop structure discussed, another example is the message exchanged when an action flows from the system to an external actor or vice versa; this again, is the subject of UML Sequence diagrams, which will be one of the topics of the next chapter.

Exercise 8.3.6: Modelling vs Architecting

For the "Online Shopping" Activity diagram created in the previous exercises:

1. What is the necessary pre-condition for this process? What if a select item is out-of-stock?
2. What is the post-condition for this process? Can we assume that the customer will always achieve this condition?
3. What is missing based on your experience and knowledge of using an OSS? Are there any other flows? What about actors other than the Customer?

In addition to just modelling, these questions start to lead into system architecting that is beyond system modelling.

8.4   UML FOR HOLISTIC ARCHITECTURE SPECIFICATION

Although SysML is specifically aimed at systems engineering, making it seemingly a better language for system modelling, UML is favoured in this book for holistic architecture specification for the following reasons:

- UML is fundamental to all modelling languages that are profiles of UML. As previously mentioned, SysML, is a UML (specifically UML 2.0) profile. Although this does not mean that one must learn UML to understand SysML, knowing how UML works can make it easier to learn SysML. UML is also essential in understanding many other profiles, such as the UML profile for MARTE (OMG 2019), which is very well-known for the modelling of real-time and embedded systems.
- Compared with SysML, the use of UML will not be disadvantaged in the application of the architecting process presented in this book. SysML has significant overlaps with (as well as differences from) UML. The overlaps and the differences are often explained at the diagram level rather than the model element level, as previously discussed in Chapter 4, Section 4.1.2. Apart from the Class diagram, there is no real difference in whether UML or SysML is used to facilitate the architecting process presented in this

book. The authors acknowledge that SysML Blocks are more suitable to model physical elements while UML Classes are more suitable to model software elements. This is why key diagrams offered by SysML are explained in Chapter 10 for the modelling of physical elements, particularly with the involvement of item flows between these elements through interfaces. Despite the differences observed in Classes and Blocks, the underlying modelling concept, *Classification*, is the same. This concept will be further explained in Chapter 9.

• SysML is currently going through a major revision. The SysML version 2.0 Request for Proposal was issued by the OMG on the 8 December 2017. The new version under development is considered to be a major leap forward to better facilitate model-based systems engineering. Therefore, significant changes in the SysML metamodel are anticipated, making SysML 1.5 (the latest version) less appropriate to start with.

NOTES

1. The usage of UML in this book is based on the latest version, UML 2.6.1.
2. This is often modelled by UML standard stereotype «Subsystem», as depicted in Figure 8.1.

9 Modelling Languages II
Structure and Interoperation

KEY CONCEPTS

Modelling languages
Modelling of
> *Structural decomposition*
> *Message exchange*
Interaction modelling

This chapter continues the journey of Unified Modeling Language (UML) learning with the UML Class diagram and Sequence diagram.

In Chapter 8, Section 8.3, the application of activity partitioning demonstrated in the ATM example leads to an important systems engineering concept, *functional allocation*. The allocation strategy for the system of interest addresses the question: what delivers which functionality? Here, the 'what' might be a physical entity, a software component, or even a person. Section 9.1 explains how to model these elements using UML Classes; whilst Section 9.2 discusses how to model the interaction and interoperation among these elements using UML Sequences.

As briefly highlighted in Chapter 8, Section 8.4, despite the various modifications introduced by SysML, UML Classes and SysML Blocks are nonetheless based on the same set of underlying concepts: classification based on common features and instantiation to create objects. These underlying concepts will be explained throughout this chapter. Together with the modelling concepts and architecting process learned in the early chapters, these concepts will be explored in further depth in the last part of this chapter.

The practice of modelling languages for systems architecting presented in this chapter is therefore organised as follows:

Class Diagram
Sequence Diagram
Modelling vs Architecting

9.1 CLASS DIAGRAM

In UML, a Class diagram is a type of structural diagram that uses *Classes* and their relationships to represent the static structure of the modelled subject (the system and system elements). Arguably, the Class diagram is the most important and useful diagram for

DOI: 10.1201/9781003213635-9

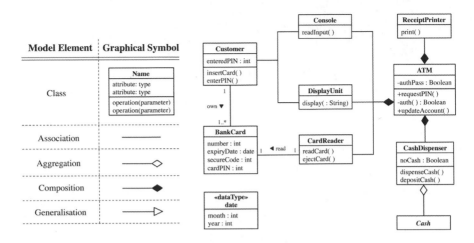

FIGURE 9.1   UML Class diagram notations and example: ATM Elements

object-oriented software programming, as coding in this fashion is centred on creating interrelated software classes. For the more general systems engineering, Class diagram, though potentially limited by its original scope, is equally important and useful due to one of its powerful underlying modelling concepts, *classification* using *Classifiers*. This is fundamental to the modelling of structure, and hence, also to architecting.

When Class, Classifier, and classification are put in one place, the interpretation of what they mean exactly can be somewhat confusing. In a nutshell, a Class is a type of a Classifier, and a Classifier represents a way of classification. There are other types of Classifiers in UML, e.g., a UseCase. To distinguish, it is easier to think of a Class as a 'template' for structural classification; whereas a UseCase is a 'template' for behavioural classification. From this illustration, it is evident that Classifier is an abstract modelling concept. It is indispensable in the construct of UML but can be suppressed when describing concrete examples in real life. For instance, in organology, *string instrument* is a human-defined class to represent one way of classifying musical instruments.

In this example of classification of musical instruments, the classification is established based on commonalities identified among different musical instruments: how the instrument utilises a common method of vibration to produce sound, e.g., instruments that produce sound from vibrating strings are classified as *string instruments*, whereas instruments that produce sound from vibrating player's lips are classified as *brass instruments*. Depending on what and how commonalities are observed and defined, different ways of classification may exist. For instance, ancient Chinese music classifies musical instruments based on the materials they are made of. Classification with Class(-ifier) in UML follows a similar construct. It is based on a set of common *Features* that can be: (i) *Properties* which are described by *Attributes* and (ii) *Operations* which can be implemented by *Methods*. These modelling concepts will be explained in detail throughout this part of the chapter with the exemplar "ATM Elements" Class diagram depicted in Figure 9.1. A list of commonly used model elements in a Class diagram and their graphical notations are also provided in the figure for referencing.

Exercise 9.1.1: UML Class Diagram

1. Under the UML project created in Exercise 8.1.1, create a new Class diagram.
2. Name the Class diagram as "Learning Classes".

Within the specific context of systems (to include software) engineering, a Class, can be defined as an abstraction of a system or system elements that share the same Features, which represent *"structural and behavioural characteristics of Classifiers"* as defined in the UML standard (OMG 2017b, 107). Graphically, as illustrated in Figure 9.1, a Class is represented by a solid rectangle in a Class diagram. It can have up to three compartments with the top compartment showing the name of the Class, the middle compartment showing the Attributes and the bottom compartment showing the Operations. The top compartment is the only mandatory compartment. The name of the class should always be centred, in boldface, and using the PascalCase. Attributes and Operations are non-mandatory and are modelled where appropriate. They are always left-justified and using the camelCase. In the "ATM Elements" Class diagram in the figure, nine classes are being specified, some of which only have attributes, e.g., the "Card" class, some of which only have operations, e.g., "Console" and "Card Reader" classes, some of which have both attributes and operations, e.g., the "CashDispenser" and an abstract class, *"Cash"* that has neither attributes nor operations. We will come back to what is meant by an abstract class later.

9.1.1 ATTRIBUTES

In UML, an attribute owned by a Class represents a *StructuralFeature*, i.e., a structural characteristic, of that class. Ideally, the specification of an attribute should constitute and reflect the rationale of the classification. Like in the ancient Chinese way of classifying musical instruments, "material" would be a suitable specification of an attribute. However, in the practice of systems engineering, attributes are naturally defined based on important engineering characteristics of the system element being modelled. For example, for a car wheel, one would define an attribute of "radius" rather than "shape". Although "shape" is seemingly more intuitive for classifying different types of wheels, "radius" is a more meaningful engineering characteristic that defines the properties of the wheels. Therefore, in the modelling and architecting process with UML, we focus on defining attributes that are meaningful in the engineering of the system. This idea is also illustrated in the "ATM Elements" Class diagram in Figure 9.1, wherein using a standard ATM, the card reader would read information such as card number, expiration date, security code, and PIN stored on the bank card. Therefore, these properties of a bank card are modelled as attributes owned by the "BankCard" class. Other characteristics such as dimensions may be important properties in the design of the card, but are not concerned by the ATM which is the focus of the system model and the Class diagram. Therefore, some characteristics may not be modelled.

In the complete specification of an attribute, the name of the attribute is followed by a colon and a *DataType*. A DataType, following an attribute, models how the value of the attribute is presented. For example, the (card) "number" of a "BankCard" has 16 digits, which is an integer, hence is modelled by the "Int" datatype. The "noCash" attribute owned by the "CashDispenser" models whether there is remaining cash in the ATM, and only has the True or the False value. Therefore, it is modelled by the "Boolean" datatype. Correct specification of datatypes of attributes is critical in software engineering due to how data is stored and read in a computer. An inappropriately defined data type could cause catastrophic failure of safety-critical systems potentially leading to loss of the system, such as the Ariane 5 accident (Lions et al. 1996). For the modelling of physical elements, the value of an attribute may have a complex structure that cannot be modelled by a single primitive type such as Integer and Boolean. In such a case, one would need to define the structure of this complex data using the stereotype, «dataType». For example, the expiry date of a bank card may be stored in two parts (month and year) rather than a single value. This is achieved by the defined "date" data type in the ATM example shown in Figure 9.1. It is interesting to note that the defined datatype, "date", has a similar graphical appearance to a Class. Indeed, just like Class and UseCase, DataType is also a kind of Classifier in which its structure is defined by its attributes, e.g., "month" and "year".

9.1.2 Operations

In addition to structural properties, a system or system element, modelled by a Class, may also have BehavioralFeature, i.e., behavioural characteristics, that can be modelled by UML Operations. Similar to the arguments put forward for the specification of attributes, in model-based systems engineering, the specification of an operation does not necessarily need to include operations that distinguish the class from other classifications. Instead, it is better to focus on *functions* that are essential in the context of the system operation. For instance, for a receipt printer to be distinguished from other types of printers, a printing function, as modelled by the "print ( )" operation, is certainly insufficient. However, in the context of the "ATM", additional details are considered unnecessary at this level of specification.

In the specification of an Operation, it is essential to describe the name of the operation using a verb or a verb-noun phrase. This name is followed by a pair of brackets, ( ), which may contain input parameters that need to be consumed to perform the operation. Once such parameters are specified, there might be the need to specify more detailed behaviours of this operation to illustrate how the operation actually works. This detailed behaviour is referred to as a *Method* in UML. The method essentially implements the operation so that the input parameters can be successfully 'converted' into output parameter(s).

An example of a fully specified authentication function ("auth ( )" operation) of the "ATM" Class in the ATM example might be:

$$auth (enteredPIN, cardPIN) : Boolean$$

which could be implemented by an algorithm (a Method) that compares the PIN entered by the customer "enteredPIN: int" and the PIN stored on the chip of the card read by the card reader "cardPIN: int". The algorithm produces one output parameter, "authPass: Boolean", which is stored as an attribute owned by the "ATM". The above Method explains how the "auth ( )" operation works in detail and this starts moving into the detailed design of the system, i.e., design of the algorithm. This is beyond architecture specification but is constrained by the architecture specification, i.e., the inputs, the output, and the allocation of the authentication function.

Finally, in the modelling of attributes and operations, as shown in the "ATM" class, there can be an additional symbol in front of the description. This symbol is called *Visibility*, which defines the visibility of the associated model element by other classes. When a visibility is not shown graphically, it means that it is either not defined or is suppressed. Visibility is an important feature for secure software coding. However, for general systems engineering, it often does not make much sense to include visibility for attributes and operations owned by physical elements. Therefore, the only class where visibilities are modelled is the "ATM" class, which is assumed to be a software element. Here, a "+" represents a *public* visibility, meaning that it is visible to any elements that can access the "ATM" class; a "-" represents a private visibility, meaning that it is only visible to the "ATM" class itself. In the rest of this book, unless explicitly informed otherwise, visibility will not be modelled to avoid unnecessary complications.

Exercise 9.1.2: Classes

Within the "Learning Classes" Class diagram created earlier,

1. Create a class and name it as "System".
2. In this "System" class, add the attribute compartment and create a private attribute with the description "serialNumber" and data type "Integer". The description in the compartment should read "-serialNumber: Integer".
3. Then add the operation compartment and create a public operation with the description "doThis ( )". The description in the compartment should read "+doThis ( )".

Having understood the basics of a Class, create a new Class diagram under the "Learning UML" project and name the diagram, "Online Shopping". In this Class diagram,

4. Create a "Customer" class with the following String-type attributes: "emailAddress", "password", and "shippingAddress"; and the following operations: "viewProduct ( )", "addItem ( )", "makePayment ( )".
5. Create an "OSS" class without any attribute and operation.
6. Create a "ShoppingCart" class with the following operations: "updateCart ( )", "calculateTotal ( )", and "generateOrder ( )".
7. Create an "Item" class with a Real-type attribute: "unitPrice" and an Integer-type attribute: "quantity".
8. Create an "Order" class with a String-type attribute: "status" and an Integer-type attribute "id".
9. Create a "Payment" class with two private Integer-type attributes, "orderID", and "cardNumber", and an attribute, "cardType" with a non-primitive type, "CardKind"; and the following two operations: "requestPayment ( )" and "confirmPayment ( )".
10. Finally, the "CardKind" data type needs to be defined through an «enumeration» stereotype. Similar to «dataType», this is achieved by defining a Class with the top compartment including the "«enumeration»" header

and the name of the data type, "CardKind" in the next line. In the next compartment, a list of enumeration literals is specified one to a line. Here, the list includes: "Debit Card" and "Credit Card".

9.1.3 ASSOCIATIONS

The relationship between two Classes is modelled by a binary *Association*. Association was introduced in the previous chapter when it was used to associate a UseCase and an Actor to model an interaction between an element in the system environment and the system. Semantically, the Association can generally be interpreted as a specification of how the Actor *uses* the system. In Class diagrams, however, the exact meaning of an Association can vary and this can be explicitly captured by its 'name', i.e., a succinct description of the meaning of the association, shown in the middle along with the Association. An arrow can also be added to the name to further clarify the direction of the relationship. For instance, in the ATM example illustrated in Figure 9.1, the "read" association between "CardReader" and "BankCard" models the relationship in which a card reader *reads* a BankCard (with the arrow pointing to the "BankCard").

On this "read" association, there is a specification of "1" at each end of the association. This is called *Multiplicity*, which models the number of the corresponding elements involved in the relationships. In this example, we have exactly one "CardReader" and one "BankCard" involved in this relationship since there is only one card reader per ATM and a card reader can only read one card at a time. In other cases, a relationship can involve multiple numbers of elements, and as such can be represented by the exact number. Very often in systems architecting, details of the exact number of involved elements are unclear until reaching the detailed design stage, therefore, the multiplicity "1.*", which represents "at least one" is often used. An example of this is illustrated by the association between the "Customer" and the "Console", modelling the case where there is at least one customer involved in this relationship. When "*" is used alone, it represents any possible (non-negative) number of involved elements and this includes 0.

UML also provides Associations for the modelling of relationships that describe a *whole-part* concept where one class, representing a *part*, is a part of another class that is representing the *whole*. For example, an ATM system as a whole consists of the following parts: a console, a display unit, a card reader and a receipt printer. UML offers two specific types of Associations, namely, *Composite Aggregation* and *Shared Aggregation* to be used for the modelling of a whole-part relationship. The name 'Aggregation' symbolises the modelling concept where the 'whole' is aggregated from the 'parts'.

Composite Aggregation, more commonly known as *Composition*, is graphically noted by a solid line with a filled diamond attaching to the end of the owning Class. It models a *strong* whole-part relationship, which is often explained in the following way: if the whole is removed from the model, the parts are also removed since their existence relies on the existence of the whole; and if one of the parts is removed from the model, the whole is considered incomplete.

Shared Aggregation, more commonly known as *Aggregation*, is graphically noted by a solid line with a hollow diamond attaching to the end of the owning Class. Compared with Composition, it models a *weaker* whole-part relationship, where the existence of the part is independent of the existence of the whole. The name 'Share' indicates that the part may be involved in multiple aggregations simultaneously to form different 'wholes'. In the ATM example, the relationship between the "CashDispenser" class and the "Cash" abstract class is modelled by an Aggregation. Clearly, the "CashDispenser" must contain "Cash", but "Cash" does not have to be owned by the "CashDispenser" only in this context, e.g., when it is dispensed, it is no longer part of the dispenser.

Exercise 9.1.3: Relationships between Classes

Within the "Learning Classes" Class diagram created earlier,

1. Create two new classes and name them as "ComponentA" and "ComponentB", respectively.
2. Connect the "System" class and the "ComponentA" class using a composition to show that the System owns Component A. Hint: the filled black diamond is attached to the "System" class.
3. Connect the "System" class and the "ComponentB" class using an aggregation to show that the System owns Component B, but Component B can exist independently of the System. Hint: the empty diamond is attached to the "System" class.
4. Connect the "ComponentA" class and the "ComponentB" class using an association to show that Component A controls (one or more) Component B(s), i.e., the description of the association reads "control". As such, the multiplicity on the "ComponentA" end of the "read" association is specified as "1" and the multiplicity on the "ComponentB" end of the association is specified as "1.*".
5. Come up with a real-world example that could fit into this Class diagram.

Having understood how to model the relationships between Classes introduced so far, continue with and complete the "Online Shopping" Class diagram by modelling the following relationships:

6. The OSS owns the shopping cart, order, and payment. These ownerships are modelled by compositions where the filled black diamond is attached to the "OSS" class.
7. Both the OSS and the shopping cart own items. As such, these ownerships are modelled by aggregations where the empty diamond is attached to the "OSS" class and the "Shopping Cart" class. Multiplicities at the "item" end are "1…*" for both ownerships.
8. The customer views and modifies the shopping cart. This relationship is modelled by an association with the name "view & modify" and multiplicities of "1" on both ends.
9. The shopping cart generates the order. This relationship is modelled by a "generate" association with multiplicities of "1" on both ends.
10. The order requests payment. This relationship is modelled by a "request" association with multiplicities of "1" on both ends.
11. And finally, the customer makes payment. This relationship is modelled by an association with the name "make" and multiplicities of "1" on both ends.

9.1.4 GENERALISATION

In addition to Association, two classes can also be related by a Generalisation. The concept of Generalisation was already introduced and explained in Chapter 8 where one UseCase can be generalised by another UseCase and an Actor can be generalised by another Actor. From the earlier illustration on how Class is related to Classifier, and knowing how UseCase, Class, and Actors are just different types of Classifiers, it is not difficult to understand how Generalisation is defined using Classifier as *"a taxonomic relationship between a more general Classifier and a more specific Classifier"* in the UML standard (OMG 2017b, 138). Essentially, Generalisation in UML describes a hierarchical structure in the modelling elements. This is very natural to how classification works.

Once more, let us use the musical instrument example for illustration. In the commonly accepted way of classification, musical instruments can be classified into string instruments, brass instruments, etc. To model this classification using UML, we would have one class defined for each type of musical instruments, for example, a "StringInstrument" class, a "BrassInstrument" class, etc. These classes are then *generalised* into the more generalised concept, musical instruments which can be modelled by a "MusicalInstrument" class. Following the same line of thought, there are also other types of 'instruments' that are non-musical, which make musical instrument itself a class to represent a classification of 'instruments'. As such, along with the "MusicalInstrument" class, there can be a "MeasuringInstrument" class representing devices for the purpose of making physical measurements, a "MedicalInstrument" class for devices used in general medicine, so forth and so on. These classes are referred to as the *specialised* classes or *child* classes of the *parent* class, "Instrument". Similarly, in the other direction, one can define lower-level classifications under, e.g., the "StringInstrument" class, such as a "Guitar" class, a "Violin" class, etc.

Viewing the example above holistically, one immediately observes a hierarchical structure of classes representing the hierarchy of classifications. The modelling of generalisation hierarchy here is not to be confused with the modelling of system hierarchy where Aggregation and Composition are used, e.g., the ATM system illustrated in Figure 9.1. We will come back to this subtle point in the last part of this chapter to understand what this means for structural decomposition in architecting.

9.1.5 INSTANTIATION

So far, classes have been used to model real-world abstraction rather than real-world objects. For instance, in the ATM example, the "ATM" represents an abstract concept of ATM rather than specifically referring to a particular ATM in a bank or on a street. Indeed, based on the definition of UML Class and Classifier, Classes were never meant to model a real-world object. The modelling of a specific object in UML is achieved through the concept known as *instantiation*. An instantiation of a class (or a block in SysML) simply means the 'creation' of an *object*. An object is also referred to as an *instance* of the class which it instantiates. When a class cannot be instantiated directly, it is referred to as an *abstract* class, for example, the "Cash" abstract class as modelled in the ATM example.

It is important to note that UML does not define object as a modelling element. Rather, to specify the details of this object, it uses *InstanceSpecification*. The

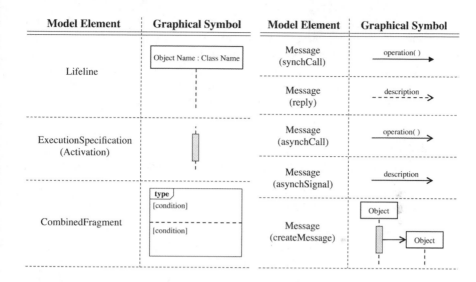

Model Element	Graphical Symbol	Model Element	Graphical Symbol
Lifeline	Object Name : Class Name	Message (synchCall)	operation( ) →
		Message (reply)	description --→
ExecutionSpecification (Activation)		Message (asynchCall)	operation( ) →
		Message (asynchSignal)	description →
CombinedFragment	type [condition] [condition]	Message (createMessage)	Object / Object

FIGURE 9.2 UML Sequence diagram notations

graphical notation of an object and examples of objects presented in a Sequence diagram is shown in Figures 9.2 and 9.3, respectively. Specifically, following the graphical notation of a Class, an object has its own name followed by a colon and then the name of the class it instantiates. When the name of the object is in absence, such an object is often referred to an *anonymous* object.

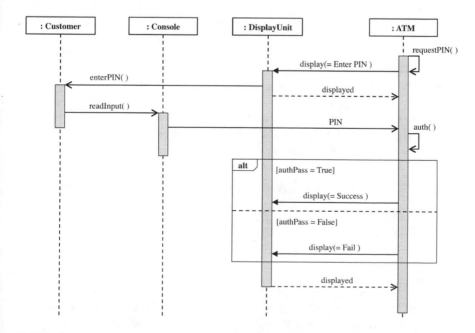

FIGURE 9.3 UML Sequence diagram example: ATM User Authentication

At the architectural level, it is rare to model a detailed design specification for a particular object (representing a specific system element), e.g., its exact material, dimension, and weight. This aligns with how the scope of the Class diagram is defined: internal to the system and stops at the class level. Hence, instantiations are not illustrated in any Class diagram, but are shown in other diagrams such as UML Component diagram, which will not be covered by this book. Although being out of scope in the construction of a Class diagram, instantiation without detailed specification is still necessary to continue the modelling and architecting of the system. One reason for this is to enable the modelling of interaction and interoperation between system elements using UML Sequences, which will be explained and discussed next.

9.2 SEQUENCE DIAGRAM

In the explanation of using multiplicities to model the potential number of elements involved in an interaction (as modelled by Associations), each individual member, i.e., an object, involved in the defined association may have its own detailed interaction mechanism. For instance, a specific customer object, "Alice: Customer", instantiated from the "Customer" class in the ATM example shown in Figure 9.1, could interact with a specific console, "test: Console", instantiated from the "Console" class to illustrate the communication required to achieve a "Display Balance" use case; whilst another specific customer object, "Bob: Customer" could interact with the same console to illustrate the detailed interaction and interoperation for the "Withdrawal Cash" use case. In this part of the chapter, we will look at how these detailed interactions are modelled using model elements in UML and presented in a UML Sequence diagram.

A UML Sequence diagram is a type of UML Interaction diagram that specifies how elements interact in detail. Other kinds of Interaction diagram include, for example, the State Machine diagram and Timing diagram. Each provides a different view in representing how the interaction and interoperation among system elements works. UML Sequence diagram focuses on the modelling of communications between objects through *Message* exchange. This includes two specific perspectives: (i) the order in which messages are exchanged to ensure desired system behaviour and (ii) the type of the messages to support interface definition. A list of commonly used model elements in a Sequence diagram and their graphical notations are also provided in the figure for referencing.

Exercise 9.2.1: UML Sequence Diagram

1. Under the "Learning UML" project created in the early exercises, create a Sequence diagram.
2. Name the diagram "Learning Interactions".

To model the exchanges of Messages between Objects, UML Sequence diagrams use a model element called *Lifeline*. A Lifeline, as shown in Figure 9.2, is graphically depicted as an object having a vertical dashed 'tail'. It models a participant involved in an interaction, specifically, in the form of communication via sending and receiving messages. It is worth emphasising that at the instance level, each Lifeline represents a single element.

The vertical line represents a timeline illustrating the process in which how this object interacts with other objects, with the time flowing downward. Nonetheless, it is important to note that such a timeline does not illustrate the exact time or time interval of interaction occurrence; it only represents the order of interaction occurrence.

Exercise 9.2.2: Lifelines

Before creating a lifeline, first, create an anonymous instance for each of the "ComponentA" "ComponentB" classes previously modelled in Exercise 9.1.3. If a professional modelling tool is used, this can often be achieved by simply dragging the model element, e.g., the "ComponentA" class, as appeared in the model element repository, to the "Learning Interactions" Sequence diagram. This should automatically create an object named ": ComponentA" and its corresponding lifeline appearing in the Sequence diagram. Otherwise, within the "Learning Interactions" Sequence diagram created earlier,

1. Create two anonymous lifelines: ": ComponentA" lifeline and a ": ComponentB" lifeline. Note an anonymous lifeline simply represents the lifeline of an anonymous object.
2. Create an "Alice: SystemUser" lifeline. If a professional modelling tool is used, observe how the tool automatically creates a "SystemUser" class in the model element repository. Otherwise, creates a "SystemUser" class and an "Alice: SystemUser" object to ensure validity of the model.

Having understood how Lifelines can be created and their relation to their corresponding Classes, create a new Sequence diagram and name it "Updating Shopping Cart". In this diagram,

3. Create an anonymous object and its lifeline for the following classes created in Exercise 9.1.3: "Customer", "ShoppingCart", and "Order".
4. Create two objects and their lifelines for the "Item" Class, with one named "toAdd: Item" and the other "toRemove: Item".

Message in UML is used to model communication as a kind of interaction between two objects presented in lifelines. As a model element that is used to model interactions between system elements, like UML Association, a Message owns a 'name' that describes the communication and specifies the sender and recipient of this Message. Accompanied with sending or receiving a Message, the participating Lifelines are modified by the creation of *ExecutionSpecification*, commonly and informally known as *Activations*, graphically illustrated by narrow rectangular boxes on top of the dashed tails of the Lifelines. This ensures the semantics in which an Object must have been "activated" before or at the moment that it starts sending a Message. Similarly, a recipient is 'activated' as soon as it receives a Message. The detailed specification of the occurrence of the ExecutionSpecification will not be considered in this book for architecture specification.

A Message has a signature that can either be an Operation or a *Signal*. When the signature is an Operation, it means that the message is meant to initiate an operation.

Such a message is often referred to as a *call*. More specifically, UML defines the *synchCall*, i.e., a synchronous call, to model a message from the sender that requests an operation to be executed by the recipient whilst the sender waits for a response from the recipient. The response, which is modelled by another kind of UML Message, named *reply*, occurs later and goes from the recipient back to the sender in this case. As illustrated in Figure 9.2, the notation for the synchCall Message is a solid line with a filled arrowhead pointing to the recipient of the message, and the notation for the reply Message is a dashed line with an open arrowhead pointing to the recipient of the message (sender of the syncCall Message). This type of communication is very commonly used in network architecture, for instance, the client-server style architecture where a client sends a request of a service from the server and the server provides something in response. The form of the interaction in this case is through the exchange of messages.

A Message can also be asynchronous in that the sender of the message does not wait for a response and continues its execution. Specifically, when an asynchronous Message has an Operation signature, it is referred to as an *asynchCall*, and when an asynchronous Message has a Signal signature, it is referred to as an *asynchSignal*. Both are graphically noted by a solid line with an open arrowhead pointing to the recipient of the Message but distinguished by the syntax of the 'name' of the Messages, as shown in the figure. On the one hand, for *asynchSignal*, the 'name', appearing on top of the Message, briefly describes the contents of the signal and this is ideally verbalised by a noun phrase, e.g., "PIN". On the other hand, for both *synchCall* and *asynchCall*, the 'name' is exactly the Operation in which the Message is calling, hence, it follows a verb phrase followed with a pair of bracket '( )', e.g., "display ( )" and "readInput ( )". This is a very important modelling rule that ensures model consistency, which will be visited in more detail in Exercise 9.2.3 later.

The specification of UML Messages, as described above, is very much software-oriented in that it assumes communications between elements are information-based. However, for the more general systems engineering, interactions between elements are not always in the form of information. For instance, it can be in the form of energy when it comes to electrical systems such as powering a high-speed train from overhead supply lines; or in the form of materials when it comes to mechanical systems such as feeding petrol from a fuel tank to a car engine. Therefore, being able to adapt the modelling concept of UML Messages to enable the modelling of other kind of interactions is essential to system architecture. We will revisit this in Section 9.3.

Finally, Messages must be either drawn horizontally or with its arrowhead pointing downward to not violate the semantics of the Lifelines in which time flows downward as introduced earlier.

Exercise 9.2.3: Exchange and Execution

Within the "Learning Interactions" Sequence diagram created earlier,

1. Create an activation for each of the three lifelines: "Alice: SystemUser", ": ComponentA" and ": ComponentB". Note that typical professional modelling tool will automatically create activations when a message is created between two lifelines.

2. Create an asynchronous message, named "this is a message", going from the very beginning of the activation on "Alice: SystemUser" to the very beginning of the activation on ": ComponentA".

3. At a position slightly lower than where the first message ends, create a second synchronous message, named "call ( )" going from ": ComponentA" to the very beginning of the activation on ": ComponentB". Note that since a message/call can only be horizontal or pointing downward, the starting points of the activation on the ": ComponentB" lifeline should be lower and horizontally aligned with the starting point of this synchronous call. Use the example in Figure 9.3 for reference.

4. At the position where the "call ( )" ends, creates a self-message, named "execute ( )" on the activation on ": ComponentB".

5. At a position slightly lower than where "execute ( )" ends, creates a reply message, named "completed" that goes from ": ComponentB" to ": ComponentA".

6. At a position slightly lower than where "completed" ends, create an asynchronous message, named "message received" that goes from ": ComponentA" back to "Alice: SystemUser".

7. If a professional modelling tool is used, observe the changes automatically made to the "ComponentB" class. There should be two new operations, "call ( )" and "execute ( )", being added to the Operation compartment of the "ComponentB" class. If not, add these operations to ensure consistency.

Having understood how messages and executions are represented in a Sequence diagram, in the "Updating Shopping Cart" Sequence diagram created earlier, model the following detailed interaction: A customer starts with selecting an item and wishes to add this item to an existing shopping cart (with more than one item already in the cart). The shopping cart updates itself by adding the selected item. The update is presented to the customer before the customer makes another action. Then, the customer selects an item in the shopping cart and wishes to remove this item from the shopping cart. The shopping cart updates itself by removing the selected item and then presents the updated cart to the customer. Finally, the customer is happy with the shopping and proceeds to the next step by requesting the generation of an order. To complete this model, create the following messages in order and with slight spacing in between adjacent messages:

8. A synchronous message "item to be added" from ": Customer" to "toAdd: Item";

9. An asynchronous message "add request" from "toAdd: Item" to ": ShoppingCart";

10. A self-message "updateCart ( )" on ": ShoppingCart";

11. A reply message "updated cart" from ": ShoppingCart" back to ": Customer"

Then, repeating the above pattern for the parts of removing an item as follows:

12. A synchronous message "item to be removed" from ": Customer" to "toRemove: Item";

13. An asynchronous message "remove request" from "toRemove: Item" to ": ShoppingCart";

14. A self-message "updateCart ( )" on ": ShoppingCart";

15. A reply message "updated cart" from ": ShoppingCart" back to ": Customer"

And finally, the flowing messages for completing the shopping:

16. An asynchronous message "shopping completed" from ": Customer" to ": ShoppingCart";
17. A self-message "generateOrder ( )" on ": ShoppingCart";
18. A create message (without a name) that leads to the creation of the anonymous object ": Order". (see Figure 9.2 for graphical notation)

Once the model is completed, check consistency between this Sequence diagram with the "Online Shopping" Class diagram. Are there any operations that are used in the Sequence diagram that were not captured in the created classes?

In practice, communications between system elements are rarely in a linear fashion that can be simply modelled by a series of Messages. Like Actions that can be organised in parallel flows, alternate flows and loops in a UML Activity diagram, UML offers *CombinedFragments* to model complex sequencings that might involve parallelisms, alternatives, options, iterations, so on, and so forth. These complex sequencings are referred to as *interactionOperators* in UML. Instead of going through every kind of interactionOperators, we will introduce *alt*, representing possible alternatives of behaviours, to illustrate how *CombinedFragments* work and provide the reader the opportunity to explore and learn other kinds of interactionOperators in the exercise to follow.

The InteractionOperator, alt, models alternative interactions in a way similar to alternate flows coming out of a DecisionNode in Activity diagrams; and in fact, it elaborates the alternate flows to specify more detailed behaviours of the system elements and their interactions. Essentially, it models behaviours that can be expressed using 'if...then...; elseif... then...; ...; else' statements. An example of the usage of alt to model alternate interactions is given in the Sequence diagram depicted in Figure 9.3. Depending on the outcome of the output, "authPass", of the operation "auth ( )", there are two possible choices for how ATM interacts with the display unit. Each of the choices (the part that follows a "then") is enclosed by a fragment with its condition (the part that follows the "if" or "elseif"s) specified as a guard expression, e.g., "[authPass = True]". The fragments are then combined with dividers drawn in dashed lines. The type of the interactionOperator used is specified at the top-left corner of the CombinedFragment. It is important to note that the semantics of the alt interactionOperator enforces that the modelled choices are exclusive such that only one of the fragments is 'executed' at a time. Whichever fragment is followed, the interaction flow continues with the first message that occurs at the end of the CombinedFragment. This is how exclusiveness is respected. In the example, either of the two choices continues with a response, named "displayed", from the ": DisplayUnit" back to the ": ATM". Evidently, if different choices lead to different subsequent behaviours, this simply suggests that the subsequent interactions or operations should be modelled within the corresponding fragments until all fragments will continue with the same interaction or operation.

Exercise 9.2.4: Complex Interaction with Combined Fragments

Within the "Learning Interactions" Sequence diagram created earlier, model the situation where the "execute ( )" operation is performed exactly three times before moving onto sending the "completed" replay message. This is achieved by:

1. Enclose the "execute ( )" operation on the ": ComponentB" lifeline by a CombinedFragment with a *loop* interactionOperator, i.e., a (single) fragment with the type on the top-left corner specified as "loop"; then
2. Right after where the "loop" is specified, add a round bracket containing the number three so that the entire phrase reads "loop (3)". The guard condition is left empty. Now the fragment is interpreted as iterating the contained behaviour for exactly three times.

Having understood how CombinedFragments are represented in a Sequence diagram, in the "Updating Shopping Cart" Sequence diagram created earlier, modify the interactions such that (i) the customer can only choose to add an item or remove an item at any one time; and (ii) the customer will repeat adding or removing items until the shopping is completed. The modification will involve the use of an alt and a loop with a nested control structure. This can be achieved by following the procedure below:

3. Enclose the first eight messages (including the self-messages) by a CombinedFragment with an alt interactionOperator where the first fragment contains the first four messages with a guard, "[new item wanted]", and the second fragment contains the last four messages with a guard, "[existing item unwanted]".
4. For now, both fragments contain the set of messages, "updateCart ( )" and "updated cart", suggesting common behaviours. As such, based on previous discussion, this should be moved outside of the fragments and without the repetition. As a result, the first fragment now contains "item to be added" and "add request", the second fragment contains "item to be removed" and "remove request", and the combined fragments are followed by an "updateCart ( )" self-message and an "updated cart" reply.
5. Enclose everything up to this point, i.e., the alt-combined fragments and the two messages that come after it, by a new CombinedFragment with a loop interactionOperator. As the loop will go on until shopping is completed, instead of specifying an argument for the loop like in the earlier part of this exercise, a guard, specified as "[until shopping completed]", is needed here.
6. The Sequence diagram will now have one large loop-fragment containing alt-combined fragments and two other messages. The remaining messages are unchanged.

9.3 MODELLING VERSUS ARCHITECTING

Throughout this chapter, there were a few occasions where the discussion went beyond introducing UML model elements and diagrams. This part revisits those discussions to illustrate how the practices of modelling enable various types of systems thinking for architecting.

System Definition and Decomposition – While introducing UML Class and classification, it was pointed out that the modelling of attributes and operations in systems engineering is not driven solely by the underlying rationale of classification, but should rather be driven primarily by characteristics of the class that are meaningful in the engineering context. As such, in the process of specifying a UML Class to define system elements, system architects are guided by identifying attributes and operations of the system element that contribute to the structural and behavioural properties of the system. This eventually leads to an insight about the essential architecture: system decomposition is more than just classification. It involves considerations of how properties and behaviours of system elements could be integrated to create desired emergent system-level properties. Methods for how this can be achieved systematically with semantic transformations to create UML graphical models have been introduced in Chapters 4–7.

As used in Chapters 4 to 7 and defined more formally in Annexes-3.2.3 and 3.2.4, the term (system) decomposition *defines a system hierarchy.* In retrospect, it is evident how UML constructs align nicely with this concept in two folds. Firstly, when considering how decomposition is primarily aided by classification, UML offers Generalisation, based on a hierarchical classification, to model hierarchy of classes. Secondly, as discussed earlier, decomposition is more than classification. To compensate for this, UML offers the use of Shared and Composite Aggregations to more suitably model decomposition from the perspective of structural and behavioural characteristics. These modelling elements therefore facilitate an architect to consider a decomposition that not only manages the complexity of the system (by means of hierarchical classification) but potentially reduces the complexity.

Model Extension and Adaptation – There have been a few occasions where one notices that the core model elements provided in UML might be inadequate to allow an accurate representation of real-world elements, for instance, the modelling of complex data types and communications that are not in the form of message exchanges. Nonetheless, because UML is an extensible language, we have also seen how some of the issues can be resolved by using UML stereotypes (Chapter 8) and enumerations (Exercise 9.1.2). To enable proper modelling of interactions to go beyond exchange of messages (information), for instance to include exchange of material and energy; in this book, the semantics of a Message is extended slightly to allow its description to cover any meaningful abstraction that models a real-world exchange mechanism. In addition, as shall be explained in Chapter 10, SysML Internal Block diagram provides another solution from a structural viewpoint using (item) flows between elements.

In practical modelling of complex systems, any extension of UML should be properly made through UML profiling, where additive modelling concepts, modelling elements, and notations are formally specified. It is important for the architect to realise the deficiencies of a modelling language against the architecting task at hand in order to extend the modelling language for a more accurate representation of the architecture.

Model Consistency and Concordance – A consistent and concordant set of system models offers conflict-free interpretation of the models for an accurate implementation of the architecture. Furthermore, consistent models give architects higher confidence in assuring the correctness of architecture. Consistency should be ensured on three levels: metamodel level, model level, and context level. Below some ideas with examples are provided to illustrate how model consistency can be maintained.

At the metamodel level, i.e., UML model elements that are yet to be specified, with the use of professional tools that conform to the UML standard, it is possible to check consistency between UML diagrams based on how model elements are conforming to the rules defined in the UML standard. For instance, as previously discussed, the lifeline of an object appearing in a Sequence diagram must be specified consistently with the class that it instantiates in that it only has operations that have been defined in the class.

At the model level, it is also possible to use modelled relationships to examine the consistency between UML diagrams. For example, if two classes are related by a defined association as appearing in a Class diagram, then to maintain consistency, there should be message exchange(s) between their instantiations in the corresponding Sequence diagram. For instance, in Figure 9.1, the "Customer" class is not directly associated with the "ATM" class; this is then respected in the Sequence diagram in Figure 9.3 where there is no message exchange between an "Customer" instance and an "ATM" instance.

Finally, there are also situations where specified elements and relationships in the model are not directly related. However, at the context level, these model elements may be used to describe the same property or behaviour of the system, but at different levels of representation. For example, the concept of customer authentication has appeared in the "ATM Functionality" Use Case diagram in Figure 8.1 as a use case; in the "ATM Balance Checking" Activity diagram in Figure 8.3 as an action with dedicated alternate flows; in the "ATM Elements" Class diagram in Figure 9.1 as an operation owned by the "ATM" class; and in the "ATM User Authentication Sequence" diagram as a complex interaction.

In the various approaches presented above, model consistency and concordance checking with UML is rooted in comparing two UML diagrams that are interrelated through: (i) relationships between the (model or modelled) elements and (ii) relationships between the (model or modelled) relationships.

Exercise 9.3: Model Consistency and Concordance

1. Revisit the "Updating Shopping Cart" Sequence diagram and the "Online Shopping" Class diagram created in the previous exercises. Are there any operations that were modelled in the Class diagram that should be used in this Sequence diagram but are not being used? Revise the Sequence diagram to reflect your thoughts and provide a rationale to justify your *architecting* choices.

2. In addition to the Sequence and Class diagrams, revisit the "Online Shopping" Activity diagram created in the exercises in part two of Chapter 8. Examine the consistency among the diagrams. Consider both the consistent usage of modelling elements and concordant views presented in different diagrams.

3. Finally, revisit the "OSS" Use Case diagram created in part one of Chapter 8. Which of the modelled use cases are elaborated in the other diagrams? Examine accuracy and completeness of the system model (as represented in the set of UML diagrams) in terms of whether they sufficiently elaborate the functionality to enable the next step in systems engineering, i.e., Design Definition. Use UML model elements to revise the model to reflect your proposed changes to the architecture.

In summary, from the discussion in this part of the chapter, it is evident that modelling is not equivalent to architecting but is the foundation to any model-based systems engineering approach. The central concluding remark is that Chapters 8 and 9 have introduced four types of UML diagrams and the relevant UML model elements. By no means do these two chapters give a complete picture of UML. Rather, the distilled version summarised in these chapters is meant to set a starting point that is sufficient to perform System and Architecture Definition through graphical modelling.

10 Modelling Languages III
SysML Extensions

KEY CONCEPTS

Modelling languages
Composite structure
Modelling of interfaces
Requirements traceability

In this last chapter of Part III, we will visit the extensions introduced by the Systems Modeling Language (SysML) and explain how some of the key SysML diagrams can complement UML to achieve the modelling of physical elements, interfaces, and requirements.

To aid the learning of SysML, a running example of diesel vehicle subsystems modelling for the purpose of vehicle emissions control and reduction will be used to explain the model elements and modelling concepts introduced by SysML. This is to link with the advanced driver assistance systems example introduced in Part I, bridging architecture and design through modelling. Model elements of SysML as explained in this chapter are extensions of the UML model elements explained in the previous two chapters. The exercises are formulated in a less prescriptive way compared with the previous exercises to make them more challenging. They are centred on improving the graphical models in the running example for better accuracy and completeness.

The practice of modelling languages for systems architecting presented in this chapter is organised as follows:

Block Definition Diagram
Internal Block Diagram
Requirement Diagram

10.1 BLOCK DEFINITION DIAGRAM

In a SysML Block Definition diagram, the model element that serves as the building block of the diagram is named *Block*. Technically, Blocks can be understood as a SysML version of the UML Class, because it is fundamentally based on the concept of Classifier and classification that has been explained in Chapter 9. The details in terms of specification and representation, nonetheless, have been significantly extended and modified to make the modelling of system elements more intuitive, particularly for hardware and interfaces. Graphical notations of the model elements used in a Block Definition diagram are depicted in Figure 10.1 along with a simple example of scoped vehicle subsystems provided in Figure 10.2.

DOI: 10.1201/9781003213635-10

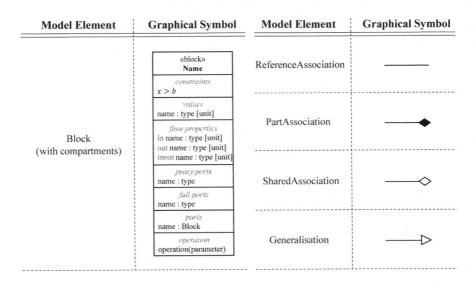

FIGURE 10.1 SysML Block Definition diagram notations

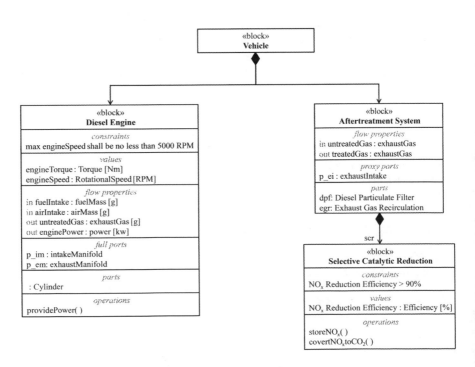

FIGURE 10.2 SysML Block Definition diagram example: Vehicle Subsystems

Exercise 10.1.1: SysML Block Definition Diagram

1. Using a preferred modelling tool, create a new SysML project with the name "Diesel Vehicle System Model".
2. Under this project, create a new Block Definition diagram with the name "Vehicle Subsystems".

As seen in Figure 10.1, compared with a UML Class, the major modification introduced for a Block is the addition of several more compartments each explained in detail as follows.

Value, being analogous to attributes of a UML Class, is used to specify measurable characteristics of the block, for example, "engineTorque: Torque [Nm]" for the "Diesel Engine" block as in Figure 10.2. The difference, however, is that value property has a much broader scope than an attribute, which primarily focuses on the specification of data types; the scope of value property is rather unbounded. In addition, it further enables the choice of unit of measurement, e.g., Newton-meter, Nm, for the *valueType*, Torque. Conveniently, OMG SysML 1.5 offers a standardised list of value types and associated units that can be used (within conforming professional modelling tools) without the need of explicit definition in the model, unlike what was needed and done in the ATM example with dataType in Figure 9.1.

ConstraintProperty, often placed in the first property compartment labelled with "*constraint*", is a special type of property that is used to capture design constraints imposed on the block, e.g., constraints as defined in the requirements or physical constraints imposed by laws of physics. The specification of these constraints should ideally follow the mathematical construct as described in Chapter 3, Section 3.3.1. As modelling tools may not always support mathematical symbols, textual descriptions of the mathematical constraints can be used as an alternative means of constraints specification. However, the underlying mathematical concept should be respected. For instance, a "max engineSpeed shall be no less than 5000 RPM" is used as a substitute for the mathematical constraint, $\max engineSpeed \geq 5000$ RPM, but the symbol '$\geq$' is often not supported.

FlowProperty, being another special type of property, is used to represent elements that flow *in* or *out* of the block, or simply the inputs and outputs. In the example in Figure 10.2, the diesel engine consumes fuel and air as input to produce power and exhaust gas, hence the specification of two in flow properties and two out flow properties, respectively, for the "Diesel Engine" block. As the exhaust gas is consumed by the aftertreatment system in the vehicle, it is then modelled as an "in exhaust gas" property in the "Aftertreatment System" block. If the output property is also the input property, like in a closed-loop control system where the output signal is feedback into the system, the *inout*[1] literal value is used to indicate the flow direction.

The specification of FlowProperties essentially requires the specification of interfaces, i.e., the system elements that handle the flows. This is achieved by using the *Port* property. At the Block level, they can be either defined as a *proxy* port or a *full* port in the corresponding compartments. The difference between the two types of ports is that a proxy port acts as a proxy for its owning block by exposing the features of the owning block; while a full port is essentially a separate system element, e.g., "em: exhaust manifold" for the "Diesel Engine" block, that is 'sitting on the edges' of the owning block and has its own features. With this special characteristic, a full port can be considered as a sub-system (sub-block) of the owning system (block) but should be differentiated from other sub-systems, as explained next, due to its special purpose, i.e., being an interface.

Last but not least, *PartProperty*[2], being one of the essential elements for the specification of the internal structure of a block in a SysML Internal Block diagram (see Section 10.2 for detailed explanation), is used to capture sub-blocks that are forming a (strong) whole-part relationship with the block. In the example provided in Figure 10.2, the vehicle aftertreatment system is in fact a combination of several technologies each responsible for the reduction of specific harmful contents in the exhaust gas. These technologies, such as exhaust gas recirculation, are treated as subsystems in this situation and modelled as part properties of the "Aftertreatment System" block.

As a final note, the modelling of block *operations*, i.e., functions owned by the block, is not different from modelling of class operations. Furthermore, the visibility of properties and operations are always public in Blocks.

Exercise 10.1.2: Blocks

1. Within the "Vehicle Subsystems" Block Definition diagram, recreate the four blocks and their properties and operations as depicted in Figure 10.2. The relationships will be the subject of the next exercise.

Like in a UML Class diagram where various relationships have been introduced to model the hierarchical structure of the classes, Internal Block diagrams utilise a similar set of relationships with slight modifications. Apart from *Generalisation*, which is essentially the same as UML Generalisation, SysML inherits the syntax and semantics of UML Association, Composition and Aggregation relationships, but instead names them as *ReferenceAssociation*, *PartAssociation*, and *SharedAssociation*, respectively.

Interestingly, with PartProperty and PartAssociation, SysML actually offers two different ways of representing the same hierarchical structure of blocks. For illustration, the part, "scr: Selective Catalytic Reduction", is explicitly presented as a sub-block owned by the "Aftertreatment System" block through the usage of a part association, rather than an owned part, like the other two parts owned by the block. The usage of the name "scr" implies the instantiation of the "Selective Catalytic Reduction" as a specific 'object' in UML speak; and note in this representation, "scr" is depicted as the property name at the part association end attaching to the "Selective Catalytic Reduction" block. This will avoid confusing a Block from its instantiations. Although diagrammatically the two methods look different, the system model as developed in any professional tool should capture such a hierarchy as soon as one of the methods is followed.

Exercise 10.1.3: Relationships between Blocks

1. Explore the two ways of representing a whole-part relationship by recreating the part associations shown in Figure 10.2. In the model element repository, observe the changes that happened to the part properties owned by the "Aftertreatment System". Confirm your observation by removing the "Selective Catalytic Reduction" block in the diagram. Test your understanding by creating an "Exhaust Gas Recirculation" block

and associate it with the "Aftertreatment System" block. When should
a part property be modelled explicitly as a block in the system model?
2. An exhaust gas recirculation technology is used in an internal combustion engine to recirculate a controllable portion of the (untreated)
exhaust gas back to the engine to reduce NOx emissions. Based on this
technology description, complete the properties and operations of the
"Exhaust Gas Recirculation" block.
3. Following the completion of the previous task, the flow properties of the
"Diesel Engine" block need to be updated accordingly. How should this
be done? What would be the problem in doing so?

Following this last question in Exercise 10.1.3, it is evident that the explicit modelling
of flow properties in a Block Definition diagram may lead to problems in how the
information should be interpreted. For instance, the appropriate revision of the flow
property "untreatedGas: exhaustGas" owned by the "Diesel Engine" block with an
"inout" literal value would raise the question: what does this really mean? Clearly,
based on domain knowledge, this is a feedback flow but achieved through the use of
the exhaust gas recirculation technology that is owned by another (sub-)system. The
diagram clearly does not illustrate such an interpretation. Using conventional associations like in the UML Class diagram also does not solve this issue completely as flow
property is primary considered as behavioural in the UML construct. This is where the
SysML Internal Block diagram comes in to offer a promising solution in modelling the
(unordered) flow structure by unpacking the detailed internal structures of the specified blocks. This will be the subject of Section 10.2 of this chapter.

10.2   INTERNAL BLOCK DIAGRAM

The purpose of an Internal Block diagram is to capture the internal structure of a
modelled block or a set of blocks, thereby unpacking the black-box system structure
into a white-box representation. The scope of concern of the diagram is more than
just system decomposition; it also provides a full picture of *what* items flow in and
out of system elements. However, it is important that this does not make an Internal
Block diagram a behaviour diagram because it does not concern *how* items flow.

SysML does not introduce a whole new set of model elements for the creation of
Internal Block diagram. Instead, Block as introduced in Section 10.1 is also the fundamental building block of an Internal Block diagram. Nonetheless, to enable the
modelling of internal structure, it utilises a very important UML modelling element,
StructuredClassifier, or more commonly known by the modelling concept and UML
diagram type, *composite structure*, which is to be explained in this part of the chapter. Figure 10.3 provides the graphical notations of the model elements (mostly introduced in Section 10.1) when they are drawn in an Internal Block diagram. An example
Internal Block diagram that builds on the previous example in Figure 10.2 is depicted
in Figure 10.4, where the internal structure of the "Vehicle" block is captured.

Exercise 10.2.1: SysML Internal Block Diagram

1. Under the SysML project created in Exercise 10.1.1, create a new Internal
Block diagram named "Vehicle Subsystem Internal Structure".

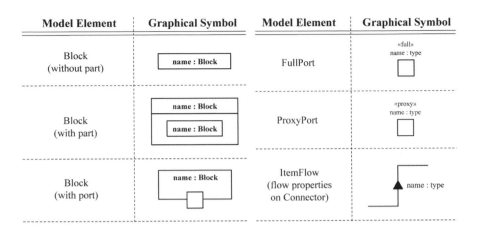

FIGURE 10.3   SysML Internal Block diagram notations

A StructuredClassifier, as defined in the UML standard, can be simply understood as a Classifier that has an internal structure comprised of linked Classifiers. Essentially, when a StructuredClassifier is fully specified, the diagram will have a nested appearance where a class may contain another instantiated lower-level class that itself may contain another next-level class. This is why the name, composite structure, is given to this way of modelling class hierarchy. StructuredClassifier also has an external structure consisting of ports. Essentially, these are interfaces in which a specified

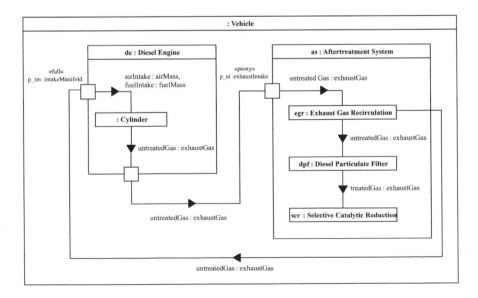

FIGURE 10.4   SysML Internal Block diagram example: Vehicle Subsystems Internal Structure

structured classifier can exchange with other structured classifiers that are not internal to it.

Analogous to a StructuredClassifier, SysML Block uses a very similar construct: all of its properties can be drawn as the internal elements in the Internal Block diagram with dedicated graphical notations and modelling rules. Details of how these properties are modelled are as follows.

Parts – As previously explained, these are sub-blocks owned by a block. As such, they form the basis of the internal structure of the owning block. Graphically, all blocks, whether it is the whole or a part in the whole-part relationships, are depicted by a rectangle box with a header in the form of "instanceName: BlockName", e.g., "de: Diesel Engine", with instance name can be left blank, i.e., an anonymous instance. Through this construct, it is evident that a key feature of the Internal Block diagram is that specified blocks are all instantiated in the diagram; thereby modelling the white-box structure at the 'object' level.

Ports – Graphically depicted as a small square, a port always sits on the edge of a block instance, as shown in Figure 10.4. The type of the port is shown by its corresponding stereotype keyword, «proxy» or «full» for proxy port and full port, respectively. As discussed previously, ports are basically interfaces owned by blocks. Therefore, connecting ports in the Internal Block diagram provides a picture on how blocks are interfaced with each other. This is achieved by using *Connectors*, which are further specialised into *BindingConnector*, *BidirectionalConnector*, and *UnidirectionalConnector*. For simplicity, this book always uses BidirectionalConnector to connect a pair of ports.

Flows – Unlike how other properties are presented in Internal Block diagrams, flow properties are modelled by *ItemFlows*, which are graphically depicted as a black arrowhead attached on a Connector, shown in Figure 10.3. The direction of the flow follows the direction the arrowhead is pointing. Therefore, the choice of direction has to be consistent with how the flow properties are modelled previously in the Block Definition diagram. For instance, as shown in the example in Figure 10.4 with a reference to Figure 10.2, the flow property "untreatedGas: exhaustGas" has an "in" literal value under the "Diesel Engine" block while an "out" literal value under the "Aftertreatment System" block. This means that the "untreatedGas: exhaustGas", when specified as an ItemFlow, needs to point from the "p_em: exhaustManifold" port owned by "Diesel Engine" to the "p_ei: exhaustIntake" port owned by "Aftertreatment System". Here, "exhaustGas" is regarded as the item conveyed.

As a final note, constraint properties and value properties, as discussed previously, can both be captured internally in a block like the part properties. The difference is that they are respectively typed by valueType and constraintBlock, whereas part properties are typed Block. These properties are more often used in SysML Parametric diagram for mathematical-based analyses. By doing so, this diagram enables traceability of model elements to these mathematical analyses. However, this type of diagram is not covered in this book due to its limitations in mathematical expressiveness and executability. Depending on required fidelity of such analyses, advanced numerical analysis tools such as MATLAB®, Mathematica, and LabVIEW are recommended.

Exercise 10.2.2: Parts, Ports, and Flows

1. Reproduce the example provided in Figure 10.4 by using the model elements created in Exercise 10.1.2 and Exercise 10.1.3. For example, instead of creating a new instantiation of the "Diesel Engine" block, the instance named "de" should be used.
2. Given the domain knowledge that the untreated exhaust gas coming out of the exhaust gas recirculation would flow into the diesel engine through the intake manifold, model this item flow.
3. In the Block Definition diagram created previously, there are two flow properties: engine power and treated exhaust gas, modelled by "out outPower: power [kw]" and "out treatedGas: exhaustGas", respectively, that are not yet modelled in the Internal Block diagram. This violates the consistency between the graphical models. Rectify this problem by creating new ports that interface with two other new vehicle subsystems: chassis and tail pipe, where the engine power is assumed to be transmitted to the chassis for further distribution to the wheels and the treated exhaust gas is to be released to the atmosphere through the tail pipe.
4. Observe in the Internal Block diagram in Figure 10.4 where the ": Cylinder" instance is deliberately unnamed. Consider that there is a requirement for a diesel engine to have six cylinders, how would you model this without the creation of six blocks in the Block Definition diagram? As a guidance, recall that everything appearing in the Internal Block diagram is an instance, or in UML speak, an 'object'. How are the changes made in this task to the Internal Block diagram reflected in the Block Definition diagram?

Although the elaboration of a Block Definition diagram into an Internal Block diagram is driven by a top-down system decomposition, as seen in the second last task of the exercise, it is clear that the decomposition can be complemented by bottom-up reverse engineering to achieve better modelling accuracy and completeness. The modelling of the six cylinders in the last task of the exercise is meant to address a particular system (implementation) requirement. In a model-based approach to systems architecting, how can we establish such traceability from the system architecture to the requirement that it addresses? This is what a SysML Requirement diagram is meant to achieve and will be the subject of Section 10.3.

10.3 REQUIREMENT DIAGRAM

The SysML Requirement diagram can be understood as a diagrammatic approach to capture textual requirements with a focus on the relationships between requirements and their traceability to the system model. As such, it is a structural diagram. In a typical project utilising conventional document-based system engineering approach, requirements of the system are usually specified purely as a textual document. This would make it difficult to establish and maintain traceability from the requirement documentation, often known as the requirement specification, to other engineering artefacts, e.g., a non-functional requirement that is satisfied by a design decision or a functional requirement that is verified by a test case definition.

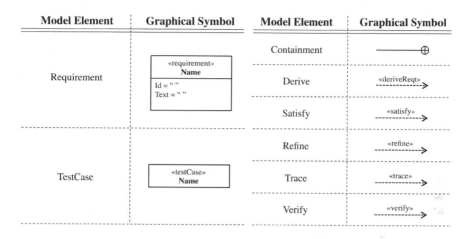

Model Element	Graphical Symbol	Model Element	Graphical Symbol
		Containment	⊕
Requirement	«requirement» **Name** Id = " " Text = " "	Derive	«deriveReqt»
		Satisfy	«satisfy»
		Refine	«refine»
TestCase	«testCase» **Name**	Trace	«trace»
		Verify	«verify»

FIGURE 10.5   SysML Requirement diagram notations

Requirement diagram is introduced in SysML with the intention of solving this problem by requesting textual requirements to be graphically specified within the system models such that traceability from requirements to model elements can be conveniently established using various pre-defined relationships. Figure 10.5 depicts the list of graphical elements used in the construction of a SysML Requirement diagram to achieve its intended purposes and with an illustrative example on vehicle emissions requirements using these model elements provided in Figure 10.6.

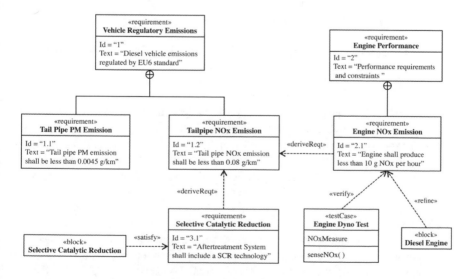

FIGURE 10.6   SysML Requirement diagram example: Vehicle Emissions Target

Exercise 10.3.1: SysML Requirement Diagram

1. Under the SysML project created in Exercise 10.1.1, create a new Requirement diagram named as "Vehicle Emissions Target".

In a Requirement diagram, each requirement statement is graphically captured within a rectangle box with two compartments. The construct is very similar to UML Class as it is essentially a stereotype of UML Class. In the top compartment, a keyword is used to show the type of requirement. By default, this is «requirement». Other useful pre-defined types of requirements in the SysML Requirement diagram extensions, as defined in Clause E.3 of SysML 1.5 (OMG 2017a, 258–260), include for example, «functionalRequirement», «performanceRequirement», «designConstraint», and «interfaceRequirement». Followed by the type of the requirement is a short 'name' of the requirement to be specified, for example, "Tailpipe NO_x Emission". In the bottom compartment, as shown in Figure 10.3, there are two default properties to be specified: the *id* property for which a unique identification number for the requirement is given and the *text* property for which the exact phrase of the textual requirement is stated. Some other additional properties are also introduced in the Requirement diagram extensions, including for example, a *source* property specifying where the requirement comes from, a *verifyMethod* property specifying the method of how the requirement shall be verified, and a *risk* property specifying the associated risks in achieving this requirement.

It has been observed that apart from the specification of the text property, the specifications of almost everything else in a requirement box are fundamentally concerned with the management of the requirements. For real-world projects where the number of requirements can be very large and may 'creep' as the project develops, proper management of requirements is demanding and often can be challenging. This is what motivates the usage of the extensions in addition to the mandatory properties.

Exercise 10.3.2: Requirements

1. Within the "Vehicle Emissions Target" Requirement diagram, reproduce the requirements as shown in the example in Figure 10.6. Leave the relationships for the next exercise.

Further to the modelling and managing individual requirements, as discussed, capturing and maintaining requirement traceability is another primary objective of the Requirement diagram. This is achieved by utilising the following relationships.

Containment, graphically drawn as a solid line with a containment symbol on one end, is a relationship used to model the hierarchical structure of the requirements. When a containment relationship is used between two requirements, the requirement with the containment symbol attached to it, also referred to as the source requirement, is said to contain the other requirement, also referred to as the target requirement; thereby creating a hierarchy between the two requirements. Once this relationship is established, it is also important to ensure the ids of the connected requirements are specified properly to reflect such a containment relationship. For example, as shown in Figure 10.6, the two tail pipe emission requirements on Particulate Matter (PM) and noxious gases (NOx) are tagged with ids "1.1" and "1.2" respectively to reflect the fact that they are contained by the "Vehicle Regulatory Emissions" requirement tagged with an id "1".

Derive, graphically drawn as a directed dashed arrow with a «deriveReqt» stereotype keyword, is a relationship used to model a requirement being derived from another requirement. For example, a system («designConstraint») requirement is derived from a regulatory requirement or a lower-level requirement on a subsystem is derived from a higher-level requirement on the system. Here, the target requirement which the arrowhead is pointing to is the deriving requirement and the source requirement is the derived requirement. To avoid confusing the derive relationship with containment relationship, it should be noted that deriving and derived requirements are individually meaningful requirements that should be addressed, whereas a containing requirement is often an auxiliary requirement introduced solely for the purpose of managing a group of common requirements, e.g., "Vehicle Regulatory Emissions". The text property of these containing requirements is often a description of the collection of the contained requirements, e.g., "Diesel vehicle emissions regulated by EU6 standard". The auxiliary (containing) requirement cannot be fully interpreted and satisfied without looking at the contained requirements. Nonetheless, if all of the contained requirements are satisfied, then the containing requirement is automatically satisfied. Obviously, this is not true in the case of a derived and deriving requirements pair.

Satisfy, graphically drawn as a dashed arrow with a «satisfy» stereotype keyword, is a relationship used to model a requirement is satisfied by a model element. The arrowhead points to the requirement being satisfied. For example, as shown in the figure, the implementation requirement, "Selective Catalytic Reduction", is *satisfied* by the inclusion of a "Selective Catalytic Reduction" block in the system model. From this example, the reader shall observe that a Requirement diagram can and should include model elements that are specified in the system models, e.g., a block or a use case.

Refine, graphically drawn as a dashed arrow with a «refine» stereotype keyword, is a relationship used to model a situation where a requirement might be refined by the evolution of a model element. The arrowhead points to the requirement that will be refined. The concept of requirement refinement can be understood by the fact that certain system requirements are not necessarily stable and their final definition is subject to the evolution of the architecture. This is particularly useful where lower-level requirements are not directly derived from higher-level requirements but are driven by architectural decisions. For example, although the tail pipe emission requirement 1.2 in the figure derives a requirement for engine-out emission (requirement 2.1), the actual engine emission target, "10g NO_x per hour", might be refined by how the diesel engine is designed and controlled. Therefore, a traceability where the "Diesel Engine" block *refines* the "Engine NOx Emission" requirement is specified. Having identified such a relationship earlier allows the monitoring of propagations of changes made to the architecture. This is an example of how requirement traceability is managed and maintained. Since the derive relationship can only be used to connect two requirements, refine relationship can also be used as a substitute when a requirement is believed to be derived from a model element that is not a requirement.

Trace, graphically drawn as a dashed arrow with a «trace» stereotype keyword, is a relationship used to model a situation where a requirement is related to another model element, but such a relationship cannot be appropriately identified as any of the relationships explained so far. Trace relationships are considered weak relationships, and as such, despite having an arrowhead that can point either way, the direction of the relationship has little meaning. It is advised that trace relationship is not overly used to avoid making the management of requirement traceability unnecessarily overcomplicated.

Finally, *verify*[3] graphically drawn as a dashed arrow with a «verify» stereotype keyword, is a relationship used to model a situation where a requirement is verified by a *TestCase*. The arrowhead of a verify relationship always points toward the requirement being verified.

SysML TestCase is a stereotype of UML Class with the keyword, «testCase». Within the Requirement diagram, it is meant to be a high-level description of the test method to be used to verify a particular requirement. Quite often, only the 'name' of the test case is specified and depicted. However, because it is a stereotyped Class, it can be useful to specify what is being measured or observed during the test as the attributes and the method of measurement and observation as the operations of the test case. See for example, the specification of the "Engine Dyno Test" test case in Figure 10.6. A test case can simultaneously verify multiple requirements.

The inclusion of a test case in a Requirement diagram enforces the necessity to define and specify the actual test method and procedure. This is achieved by using other UML/SysML behaviour diagrams such as an Activity diagram following an assurance viewpoint, i.e., assuring the system to achieve its intended purpose with an expected level of performance. Professional modelling tools support the direct linkage of such a test case as specified in the Requirement diagram to the actual test specified in another diagram.

With the introduction of the model elements required for the construction of a complete Requirement diagram, it is evident how Requirement diagram is a 'living' model in that it should be continuously updated with the evolution of the architecture and is only complete when assurance is also fully accounted for. The latter is essentially part of how the left-hand side and the right-hand side of the Vee model are linked from a model-based perspective as discussed earlier in Chapter 4, Section 4.1.1.

Exercise 10.3.3: Requirements Traceability

1. Within the "Vehicle Emissions Target" Requirement diagram, complete the example by also modelling the relationships as shown in Figure 10.6.
2. To further test understanding, using the existing requirement diagram as a reference structure, expand this diagram by including an "Engine PM Emission" requirement that states, "Engine shall produce less than 2 g PM per hour". How should this requirement be related to the other model elements (requirements or otherwise) already existing on the diagram?
3. Following the idea that one test case can be used to verify multiple requirements at the same time for the purpose of cost-saving, how would the "Engine Dyno Test" test case summary should be modified such that it can also be used to verify the new "Engine PM Emission" requirement?
4. The Block Definition diagram in Figure 10.2 has two constraints, one on the engine and another on the selective catalytic reduction technology. Model these constraints as new requirements in the Requirement diagram within the existing hierarchical structure. Then, create traceability using appropriate relationships from these two requirements to other model elements (e.g., requirements and/or blocks).

10.4 INTEGRATED USAGE OF UML AND SysML

Intuitively UML and SysML can be used in system modelling by using UML for software systems and SysML for hardware systems. In the Part II tutorials, it could be argued that such a distinction between the usage of the two languages clearly has its reasons, but is not necessary. A modelling approach that combines the languages at the diagram level or even at the model elements level can bring meaningful advantages. An integrated usage of modelling languages such as this would in fact go beyond UML and SysML. For example, when embedded systems and human 'systems' are part of the picture, the conjunctive usage of other UML profiles such as MARTE together with UML and SysML could make modelling much more descriptive. The reader should revisit the Chapter 7 tutorial to reflect on the usage of the Block Definition diagram and Internal Block diagram for hardware and interface modelling, and the Requirement diagram for traceability between requirements and system artefacts.

NOTES

1. In, out, and inout, are the literal values of the *FlowDirection* enumeration as defined in the SysML standard.
2. There is also the Reference property that is not explained in this book due to its infrequent usage.
3. Note that there is also another relationship, *copy*, which is not introduced due to its infrequent usage.

Part III Bibliography

Dickerson, Charles E., and Dimitri Mavris. 2013. "A brief history of models and model based systems engineering and the case for relational orientation." *IEEE Systems Journal* 7, no. 4 (December): 581–592.

Lions, Jacques-Louis, Lennart Luebeck, Jean-Luc Fauquembergue, Gilles Kahn, Wolfgang Kubbat, Stefan Levedag, Leonardo Mazzini, Didier Merle, and Colin O'Halloran. 1996. *Ariane 5 flight 501 failure report by the inquiry board.* Prepared by the Inquiry Board.

OMG (Object Management Group). 2017a. *OMG Systems Modeling Language (SysML®),* Version 1.5.

———. 2017b. *Unified Modeling Language (UML®),* Version 2.5.1.

———. 2019. *UML Profile for MARTE (MARTE),* Version 1.2.

Part IV

Case Studies for Practice and Practitioners

OVERVIEW OF THE PRACTICAL CASE STUDIES

This part consists of two practical case studies that span the spectrum from information-intensive systems to hardware-oriented systems. It offers students the opportunity to apply the knowledge and skills they have acquired from the earlier sections of this book. Educators who wish to use this book as a reference for academic instruction or professional development courses should find these case studies useful and informative. They should also read Annex A-5 (Using This Book for a One-Semester Module of Lectures) which shows how to use the case studies in conjunction with the other material in the book. The rationale for the choice of exemplar systems, a Traffic Management System of Systems in the first case study (Chapter 11) and a System of Actuator Systems in the second (Chapter 12), are as follows.

The first case study is concerned with an information-intensive cyber-physical system that involves the design of hardware, software, and interfaces (between systems in the system of systems context). As such, it aligns with the tutorial case studies in Part II, but is much more detailed in terms of system complexity, behaviour, and structure. This gives the student reader an opportunity to apply their acquired knowledge and skills to a challenging but feasible problem at the postgraduate level of study.

The second case study is a mechanical system that only involves hardware. The use of SysML in this case study provides the student an opportunity to apply modelling languages to a traditional discipline such as mechanical engineering. The problem challenges the student to explore the 'art' of architecture through the application of more advanced modelling concepts. The solution is not to be found in the refinement of technical details for the components of the system but instead in the architecture of the system and the reduction of complexity through modelling.

11 Case Study
Traffic Management System of Systems Architecture

KEY CONCEPTS

System architecture
Modelling and analysis
Transformation of structure
Specifications

This case study provides the student an opportunity to architect a Traffic Management System of Systems (TMSoS) that uses an Intelligent Ramp Metering System (IRMS) as the building block to improve traffic flow and reduce the risk of accidents. The chapter provides background information on the system to be architected, partially complete models as starting points for specific tasks, and prescriptive procedures for how to develop the architecture. In this way, the case study helps the student to develop skills and proficiency in system architecting and design.

The chapter is structured into three parts aligned to the chapters in Sections II and III. The student will apply the relevant processes and methods from Chapter 4 to model the traffic flow problem and a proposed solution. The student should also refer to the corresponding tutorial chapters in Part II for guidance as needed. UML and SysML diagrams produced for the case study should conform to the standards in the modelling language chapters (Chapters 8, 9, and 10).

The context is as follows. Imagine that you are a Systems Architect given the task to perform modelling and analysis for an urban road system traffic improvement project. Specifically, you will produce a system and architecture specification through a disciplined application of:

System Definition
Architecture Definition
Architecture Refinement and Analysis

As with the tutorial chapters in Part II, the format of the case study is that of a dialogue between the architect and the stakeholders that should lead to agreement on the specification of a solution to the problem.

DOI: 10.1201/9781003213635-11

11.1 SYSTEM DEFINITION

The first part of the case study is concerned with two essential technical processes: Requirements and System Definition. The aim is to create a *technical package for a system specification of the IRMS* based on a given description of the system concept. This technical package shall include a set of UML diagrams complemented with explanatory text detailing essential information.

11.1.1 CLASS EXERCISE AND PROCEDURE

To achieve the above aim, the following specific objectives (as tasks) are given:

- Use a professional modelling tool to produce a system specification consisting of a system
  - environmental model clearly capturing the separation of the system from the environment
  - behavioural model clearly capturing the events occurring in the system and the environment and the sequencing of the events
- Document explanatory text that support the modelling and architectural decisions, and assumptions made.
- Summarise the technical package for the system specification of the IRMS in a presentation for the acquiring stakeholder.

To complete these tasks, the following procedure can be adopted (not necessarily in a linear fashion) or tailored according to justifiable rationales:

1. If the work is to be completed by an individual, then the work should be planned appropriately; otherwise, assign a Chief Architect who should further partition and allocate the work to team members in addition to planning of the work.
2. Start with the system narrative provided in Section 11.1.3 to understand and analyse the narrative in a structured way.
3. Use the narrative and the understanding obtained from the analysis of the narrative, first to create a system environmental model detailing the separation of the system from the environment through the use of a system-level Use Case diagram. The diagram should clearly capture the IRMS system functionalities and interactions with its environment.
4. Subsequently, create a system behavioural model detailing the decomposition of the use cases into actions, the functional flow formed by the actions, and the allocation of the actions to a proposed set of subsystems. For the functional flow, focus on the basic flow only, i.e., do not consider alternative behaviours at this stage yet.
5. Evaluate all diagrams in terms of their syntactical and semantic correctness, completeness against, and traceability to the narrative, and consistency between model elements defined in different diagrams. Revise the model subject to the evaluation outcome.

6. During the completion of the above tasks, capture all assumptions and justification made to support the modelling and architecting of the system.

7. Summarise all of the artefacts created into a succinct system specification (see Chapter 5, Section 5.5, for example) and prepare a presentation for the acquiring stakeholder.

11.1.2 CLASS ASSESSMENT

The package shall be assessed through the following three viewpoints:

- Correctness and completeness of the contents. Specifically, this means
  - a correct and structured analysis of the narrative showing a complete set of derived functional requirements
  - correct usage of UML syntax and semantics in the UML diagrams created
  - a complete set of functionalities of the IRMS and interactions with the environment
  - appropriate and well-reasoned initial functional decompositions and allocation
  - a correct basic functional flow for a complete set of identified scenarios.
- Desired qualities of the system model, which are
  - traceability of the UML diagrams to narrative
  - consistency between the UML diagrams
  - explanatory text on key points, architectural decisions, and assumptions made
- Appropriateness of the overall approach to the case study, which includes
  - work planning, breakdown, and allocation
  - presenting the package to stakeholders in an appropriate format

11.1.3 SYSTEM CONCEPT DESCRIPTION

An initial project narrative, adapted from published research (Ingram et al. 2014; Dickerson et al. 2016), is given by the acquiring stakeholders as follows:

This project aims at improving UK urban road traffic by the design and implementation of a Traffic Management System of Systems (TMSoS), which is a collection of diverse systems that work together collaboratively to control road traffic. Its goals include but are not limited to improving traffic flow and reducing accident rates.

At the heart of the TMSoS are the so-called Intelligent Ramp Metering Systems (IRMS) which individually control the traffic flowing into the motorways from a local motorway access ramp. This shall be achieved by using a two-phase traffic signal to stop or admit vehicles on the ramp. The rate at which the vehicles are admitted, known as the admittance rate, depends on the phase duration employed by the IRMS. The IRMS has a default, fixed-time operational mode that changes phases at a pre-defined fixed time interval.

An IRMS shall collect local traffic data through the use of traffic sensors such as inductive-loop traffic detectors and cameras that are most suitable to local road and weather conditions where the IRMS is situated. The local data shall be continuously passed onto the regional Traffic Control Centre (TCC) to which the IRMS belongs.

A conceptual illustration of an IRMS suited on a motorway access ramp is provided in Figure 11.1.

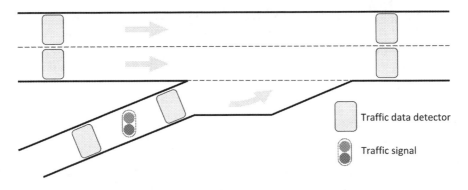

FIGURE 11.1   IRMS conceptual illustration

11.1.4   INSTRUCTOR GUIDANCE

Before developing the system specification, the student should complete the study of Part I of this book and use the first tutorial (Chapter 5) in Part II as a reference.

The first step of tackling this architecting problem is to analyse the given narrative by means of annotating it to understand what the system requirements are, functional or otherwise. It is reasonable that the student raises questions on what exactly a statement in the narrative means, similar to negotiating with the actual acquiring stakeholder in the real-world. However, it is important to realise that the purpose of this part of the case study is not to write a formal requirements specification, but to use credible sources with reasonable assumptions to practice the architecting of the system behaviour in its operating environment. Hence, the student is recommended to perform additional, but scoped research to gain further understanding of how the system is expected to work and make (and justify) reasonable assumptions where necessary.

Once the narrative is analysed and understood, the case study should move on with a combination of architecting and modelling tasks. Specifically, one possible approach is detailed as follows. First, on the basis of the annotated narrative along with any additional knowledge acquired and assumptions made, list the functionalities that the IRMS needs to achieve its operational purpose. For example, the statement, *"The local data shall be continuously passed on to the regional TCC which the IRMS belongs to"* implies that the IRMS needs a data uploading functionality. This then becomes the basis for the specification of an "Upload Traffic Data" use case in a system-level UML Use Case diagram. The identification of the associated element(s) in the environment of the system (IRMS) would be the next step. In the example of uploading traffic data, an obvious element in the environment would be the TCC. This information then enables the modelling of a "TCC" actor and its association to the "Upload Traffic Data" use case in the UML diagram. Repeating this approach for all the functionalities identified would produce an initial, yet possibly incomplete system-level UML Use Case diagram. To complete the diagram, the subsequent task is to identify the hierarchical structure among the functionalities and to connect the defined use cases with appropriate UML relationships, such as «include» and «extend» to reflect this functional hierarchy.

Essentially, the above approach follows an 'inside-out' approach where the model elements are defined first but their structure is architected subsequently. An alternative approach to the modelling of the system functionalities and its environment would be the typical 'top-down' approach where instead of identifying all functionalities, one would start with the highest level, the most abstract functionality. For the case study, the top-level functionality of the IRMS would be controlling the traffic on the ramp, which can be specified by a "Control Ramp Traffic" use case. Then, the top-down approach continues with decomposition into lower-level functionalities and modelling decomposition using relationships such as «include» and «extend». The top-down approach is arguably more intuitive to understand, but not necessarily easier to apply. This is why an inside-out approach that involves a combination of top-down and bottom-up elements is often preferred in engineering practice. There are also other approaches that adhere to different architecting styles. Whichever approach the student decides to follow, the underlying architecture principles would not differ. The resultant models following different approaches may likely differ in the graphical and textual details, but in principle should be equivalent or equally valid in terms of its *structure* (as defined in Chapter 3, Section 3.2.2).

The Use Case diagram provided in Figure 11.2 provides a starting point for the completion of the environmental model by capturing the exemplar use cases and relationships defined so far.

The next task is to create the behavioural model of the IRMS at the system level. As explained in Chapters 4 and5, this can be achieved by following a transformational process that elaborates the environmental model through adding knowledge using semantic transformation. Specifically, the knowledge to be added is the basic *functional flow*, and the resultant model will be presented in a UML Activity diagram.

In this process of transforming the Use Case diagram into a more detailed Activity diagram, the Use Case description is a useful intermediate step for capturing analysis of the use cases and capturing the interpretations. For instance, what would be the pre-condition(s) and post-condition(s) for a successful execution of the top-level "Control Ramp Traffic"? A more specific question that the Activity

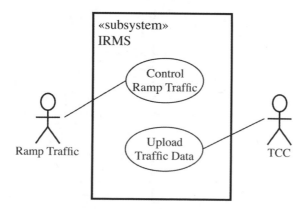

FIGURE 11.2 An initial IRMS Use Case diagram as a starting point

diagram is intended to answer is how the functionalities should be ordered. One can use both deductive and inductive reasoning to establish the basic flow. For instance, deductively, one asks what must happen just before the IRMS "Uploads Traffic Data" to the TCC? Clearly, the traffic data must be gathered, which is exactly captured by a "Collect Local Traffic Data" use case that was specified in the previous task according to the narrative. Reasoning inductively, one could ask what would happen just after the IRMS has completed the upload of the traffic data? Would it wait for an acknowledgement or would it assume that the TCC would always receive the data? Here, the answers to the questions would lead to diverse architecture solutions. There is no single correct answer at this stage because relevant information in the narrative is not provided, i.e., the narrative did not comment on how the TCC should communicate with the IRMS. A decision needs to be made based on an architectural trade-off that will analyse and compare different solutions. The justification of such a decision needs to be documented and used in the next stakeholder meeting to explain the architecture. Briefly, an 'acknowledgement' from the TCC would be more sensible for the purpose of fault diagnosis, but this could arguably result in a higher cost in the design and implementation of the IRMS with an additional functionality.

From the above example, it can be seen that in the elaboration process, not only functional flow is determined, but also additional lower-level functions may likely be introduced through the specification of additional actions. This is also known as the increasing 'design commitment' as system architecture evolves. An initial Activity diagram capturing this example is offered in Figure 11.3 as a starting point for the

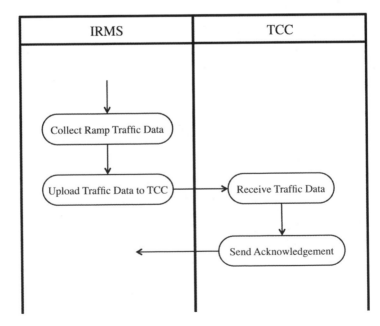

FIGURE 11.3   An initial IRMS Activity diagram as a starting point

completion of the behaviour model with the basic flow only. A control structure that involves alternate flows will be visited in the third part of the case study.

In the completion of this IRMS Activity diagram, the following questions need to be addressed:

- Is a single Activity diagram sufficient to capture all the possible scenarios which may involve different actors? If not, model additional scenarios in separate Activity diagrams. A Use Case Description table would be useful to organise your thoughts.
- Would it be possible to propose an initial set of sub-systems for the IRMS according to the Activity diagram(s) and the narrative? Allocate the defined actions to the proposed sub-systems using appropriate hierarchical partitioning. This will be an initial functional allocation.

At this point, the environmental and behavioural model of the IRMS is completed. Before continuing the architecting process, it is necessary to evaluate the models so far in terms of completeness, traceability, and consistency with respect to each other and to the project narrative. For example, one would ask the questions: are all the use cases defined in the Use Case diagram being captured and elaborated in the Activity diagram? Do all actions and partitions in the Activity diagram have a traceable origin to the Use Case diagram? Are all of the specified model elements referenceable to the narrative? If some of the model elements are additional design commitments, is there a justifiable argument for this architectural decision available for stakeholder review? The models should be revised based on answers to these questions.

Once confidence has been established in the graphical models, they can then be included in the technical package for the system specification of the IRMS. In addition to the models, the package should also include the other artefacts that have been generated during the completion of the tasks. These would likely include a document on the specified functional requirements derived from the project narrative and introduced through functional decomposition; the Use Case descriptions capturing the scenarios analysed; a traceable document that captures the justification of the architectural decisions; and an architecture evaluation on the completeness, traceability and consistency of the diagrams. Finally, an executive level summary similar to the one in Chapter 5, Section 5.5 is necessary to facilitate stakeholder meetings and communications.

11.2 ARCHITECTURE DEFINITION

The second part of the case study is concerned with the essential Architecture Definition process. The aim is to create a *technical package for an architecture specification of the IRMS* similar to the package created in the previous part, but with a focus on functional allocation and behaviour synthesis, specification of system elements and their functions, and system element interoperations. Again, this technical package shall include a set of UML diagrams complemented by explanatory text detailing essential information.

11.2.1 CLASS EXERCISE AND PROCEDURE

To achieve the above aim, the following specific objectives (as tasks) are given:

- Use a professional modelling tool to produce an architecture specification consisting of a
  - system structure model clearly capturing the specification of system elements and their properties and operations;
  - synthesised system behavioural model clearly capturing the revised functional allocation; and
  - system interaction model clearly capturing the interoperations among system elements.
- Document explanatory text that support your modelling and architectural decisions and assumptions made.
- Summarise the technical package for the architecture specification of the IRMS in a presentation for the acquiring stakeholder.

To complete these tasks, the following procedure can be adopted or tailored according to justifiable rationales:

1. If the work is to be completed by an individual, then the work should be planned appropriately; otherwise, assign a Chief Architect who should further partition and allocate the work to team members in addition to planning of the work.
2. Start with the additional system narrative provided in 11.2.3 to understand and analyse the narrative in a structured way.
3. Use both of the narratives provided so far, concurrently to
   a. create a system structure model detailing the specification of subsystems through the use of a Class diagram. The diagram should clearly capture the properties and operations required by the subsystems, the system hierarchy, and the relationships between the subsystems (according to the revised system behavioural model below); and
   b. revise the previously created system behavioural model to update the functional allocation and functional flow (according to the subsystems defined in the system structure model above).
4. Subsequently, derive and create a system interaction model detailing the specification of function calls made by subsystems and exchanges of messages between subsystems through the use of one or more Sequence diagrams.
5. Evaluate all diagrams in terms of their syntactical and semantic correctness, completeness against and traceability to the narratives, and consistency between model elements defined in different diagrams. Revise the model subject to the evaluation outcome.
6. During the completion of the above tasks, capture all assumptions and justification made to support the modelling and architecting of the system.
7. Summarise all of the artefacts created into a succinct architecture specification (see Chapter 6, Section 6.5 for an example) and prepare a presentation for the acquiring stakeholder.

11.2.2 Class Assessment

The package shall be assessed through the following three viewpoints:

- Correctness and completeness of the contents. Specifically, this would mean
  - a correct and structured analysis of the narrative showing a complete set of additionally derived system requirements for Architecture Definition;
  - correct usage of UML syntax and semantics in the UML diagrams created;
  - a complete set of subsystems of the IRMS and elements in the environment with sufficiently detailed properties and operations, hierarchy, and relationships;
  - correctly revised functional allocation; and
  - a complete set of correctly ordered execution of functions and exchange of messages.
- Desired qualities of the system model, which are
  - traceability of the UML diagrams to the narrative;
  - consistency between the UML diagrams; and
  - explanatory text on key points, architectural decisions, and assumptions made.
- Appropriateness of the overall approach to the case study, which includes
  - work planning, breakdown, and allocation; and
  - presenting the package to stakeholders in an appropriate format.

11.2.3 System Concept Description

Assuming that a meeting with the acquiring stakeholders has taken place to review the technical package completed in the previous part of the case study, the following additional requirements have been added to the existing narrative:

> To facilitate the collection of local data, the IRMS shall utilise two sets of inductive loops on the ramp with one set prior to the traffic signal and the other set after the traffic signal to allow detection of non-compliant vehicles in addition to just flow rate. Furthermore, the IRMS shall also use two sets of inductive loops or cameras on the motor way in the proximity. The first set shall be suited at an appropriate distance from the ramp to allow meaningful calculation of local motorway traffic flow while providing enough time for the processing, analysing, and uploading of the local traffic data by the IRMS. The other set shall be suited at an appropriate distance after the ramp to allow feedback to be sent back to IRMS for confirmation. All of the data gathered by the above sensors are part of the package to be uploaded to the TCC. Figure 11.1 should be used as a reference.
>
> To facilitate the implementation of intelligent control strategies, the IRMS shall have as a minimum a dedicated control unit to activate/deactivate other units and their functions.

11.2.4 INSTRUCTOR GUIDANCE

Before developing the architecture specification, the student should complete the study of the second tutorial (Chapter 6) in Part II.

The first step of tackling this architecting problem is to analyse the additional narrative to identify detailed (functional and implementation) requirements on subsystems by asking the questions: what are the system elements required? what do they need to do? and how do they collaborate to achieve the system-level purposes? It is very likely that domain-specific knowledge, e.g., how an inductive loop works, is required to appropriately architect the system. Therefore, the student is encouraged to perform additional research to gain further understanding of relevant technologies. Further reading should be again limited to generic technologies rather than specific ones to avoid making design decisions too early. For instance, wireless inductive sensors would be a technology that is too specific as compared to the generic induction concept. Using classes to specify system elements (as types of objects) should help to avoid this issue.

Once the narrative is analysed and understood, the case study should move on with a combination of architecting and modelling tasks. Specifically, one possible approach is detailed as follows. It should now be clear to the student, based on the analysis of the additional narrative, what the minimal set of system elements would be. Therefore, the first task to be completed is to revise the list of system elements and elements in the environment proposed previously to reflect both the previous and the new requirements. Furthermore, properties and operations (functions) of these elements can also be defined using existing knowledge. As such, the identified system elements can now be specified as UML Classes. As an example, the traffic signal on the ramp is required to have two phases: one for admitting vehicles and one for stopping vehicles. Therefore, the class, "Traffic Light" shall have a "signalPhase" property that takes on two possible values: "Green" or "Red"; and evidently two operations: "displaySignalPhase ( )" and "changeSignalPhase ( )". Note that an early architectural design decision has been made on 'implementing' the traffic signal by a standard Green/Red traffic light. This is to avoid ambiguity in interpreting traffic signal as the digital signal rather than the physical entity that displays the signals to the drivers. This decision is further justified by following the UK Highway Code, where a Green/Red traffic light is a standard implementation of an admittance/ stopping traffic signal. The above example provides how a UML class can be fully specified with the corresponding assumption and justification. Following the same idea, an initial UML Class diagram containing a full list of system elements, as well as environmental elements, can be established with the starting point provided in Figure 11.4. The missing information at this stage is the structure, i.e., junction and separation, which should be modelled by UML associations.

Before defining the associations, a concurrent activity that can be completed with the list of specified classes is to revise the initial functional decomposition and allocation that was done in the previous part. The outcome of this step should be an updated behavioural model with the revised UML Activity diagram that reflects the specification of system elements and corrects functional allocation where necessary. For systems that have a small number of elements, this process might be

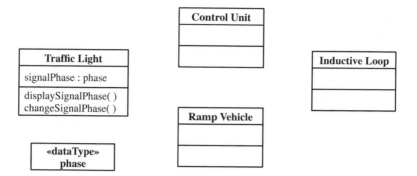

FIGURE 11.4 An initial IRMS Class diagram as a starting point

straightforward. However, with increased complexity, to ensure correctness and completeness, the student is encouraged to follow the structured methods presented in Chapter 4 with reference to their application in Chapter 6. The idea is to create new Use Case diagrams for each class of the system elements (subsystems) and model how the subsystem is proposed to interact with other subsystems and possibly elements in the environment. This is similar to treating each subsystem as a 'new' system and conducting a corresponding System Definition process. For example, the "Traffic Light" class would be entitled to the creation of a UML Use Case diagram with the «subsystem» name, "Traffic Light". Evidently, it should be interacting with the IRMS control unit to deploy traffic control strategies and also with vehicles on the ramp that are expected to follow the traffic signal deployed. This Use Case diagram is depicted in Figure 11.5.

The new Use Case diagrams then serve as the basis for the synthesis of the updated Behavioural Model. Specifically, following the same transformational approach used in the previous part, an individual Use Case diagram can be elaborated into

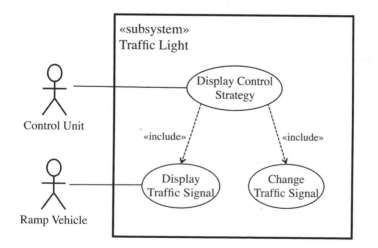

FIGURE 11.5 An example "Traffic Light" subsystem Use Case diagram

an Activity diagram that captures the intended behaviour of individual subsystems. As the Use Case diagrams are interrelated in that an actor in one diagram would be the 'subject' of one other Use Case diagram, the Activity diagrams should 'overlap' with each other in a way where flows crossing the boundaries between every two partitions can be derived from their source Activity diagrams (see Figures 6.5–6.7 in Chapter 6 for specific examples). These crossing flows are the basis for defining associations between the Classes that are previously defined in the UML Class diagram. For instance, it is anticipated that the "Traffic Light" class should be associated with an external "Ramp Vehicle" class as well as the internal "Control Unit" class, which are captured in Figure 11.4 but without properties and operations defined yet.

In the last step, with the system behaviour model presented in the synthesised Activity diagram and system structure model presented in the Class diagram, it is possible to apply the synthesis method as summarised in Chapter 6, Section 6.6 to create an interaction model capturing the interoperation of system elements presented in a UML Sequence diagram. Figure 11.6 provides a starting point. In this diagram, objects are instantiated based on the initial Class diagram provided in Figure 11.4 and the operations of the Traffic Light have been elaborated further with input arguments. The input–output relations of the operations can be used to order of execution. Some initial message exchanges between the traffic light and other elements are also provided. These exchanges trace to the associations modelled in the Use Case diagram given in Figure 11.5.

At this point, the behavioural model of the IRMS has been updated, and an initial class structure model and an interaction model are complete. Like in the previous part of the case study, before moving to the next part of the chapter, it will be necessary to evaluate the models so far in terms of completeness, correctness, and consistency with respect to each other and the project narrative. Specifically, with the new models, one would ask the questions: are all the classes correctly represented in the Activity diagram and instantiated in the Sequence diagram for the modelling of functional allocation and object lifelines respectively? Are all the messages consistent with the operations of the corresponding Class or the functional flow crossing

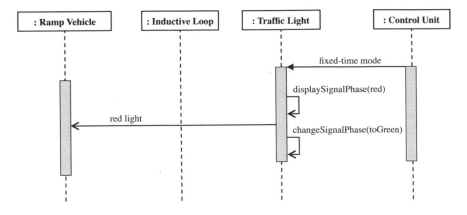

FIGURE 11.6   An initial IRMS Sequence diagram as a starting point

the partitions or both? Are they complete? i.e., do they account for all the intended interactions? Are all of the specified model elements referenceable to the narrative? If some of the new model elements are additional design commitment, is there a justifiable argument for this architectural decision available for stakeholder review? The models should be revised based on answers to these questions.

Upon having good confidence in the models, they can then be included in the technical package for the architecture specification of the IRMS. In addition to the graphical models, the package should also include the other artefacts that have been generated during the completion of the tasks. These would likely include the revised requirements document to include the new system requirements derived from the additional narrative; a document that captures the justification of the architectural decisions; and an architecture evaluation on the completeness, traceability, and consistency of the diagrams. Finally, an executive-level summary similar to the one in Chapter 6, Section 6.5 is necessary to facilitate stakeholder meetings and communications.

11.3 ARCHITECTURE REFINEMENT, ANALYSIS, AND FINAL SOLUTION

This final part of the case study is concerned with completing the essential architecture and producing an implementation model. This will be the final project solution and will enable system design in the next phase of systems engineering lifecycle. Specifically, the final project solution shall include:

1. Revised essential architecture of the IRMS to include alternative system behaviours those were temporarily and deliberately excluded in the previous parts.
2. An architecture specification of the TMSoS, i.e., a system of systems that consists of multiple IRMS and TCC, capturing collective and collaborative behaviours of systems in the system of systems (SoS) context.
3. Hardware, software, and external interface specifications of the IRMS for the design team.

11.3.1 CLASS EXERCISE AND PROCEDURE

To achieve the above aim, the following specific objectives (as tasks) are set:

- Continue with and revise the models produced previously to capture the alternative behaviours of the IRMS clearly and consistently through various UML diagrams.
- Develop further an interaction model for the SoS that clearly captures the interoperations among a set of three IRMS working collaboratively with the regional TCC as illustrated in Figure 11.7.
- Identify and define external interfaces, i.e., interfaces of an IRMS to other systems and elements in the environment in the SoS context.
- Model the interfaces in SysML Block Definition and Internal Block diagrams.

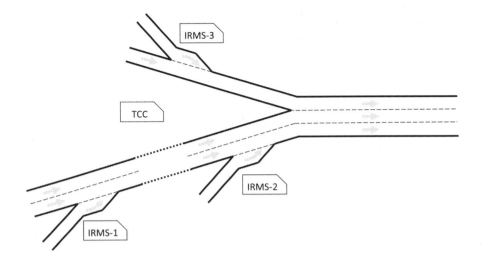

FIGURE 11.7 A system of systems with three IRMS and a regional TCC

- Perform a study to describe and analyse the critical flow of the motorway traffic and how this is affected by the admittance rate of a single ramp and multiple ramps.
- Derive a set of design constraints bounding the design space.
- Capture hardware, software, and external interface specifications in a SysML Requirements diagram with appropriate traceability to system elements.
- Define preliminary test cases that could be used to verify the requirements.
- Summarise the final solution of the IRMS and TMSoS in a presentation for the acquiring stakeholder and the design team (see Chapter 7, Section 7.4 for an example).

To complete these tasks, the following procedure can be adopted or tailored according to justifiable rationales:

1. If the work is to be completed by an individual, then the work should be planned appropriately; otherwise, assign a Chief Architect who should further partition and allocate the work to team members in addition to planning of the work.
2. Start with the additional system narrative provided in Section 11.3.3 to understand and analyse the narrative in a structured way.
3. Use the additional information provided to revise the models developed so far, which shall include:
   a. extending use cases specification in relevant Use Case diagrams;
   b. alternate paths specification in relevant Activity diagrams;
   c. additional properties and operations specification in relevant classes and in blocks per a new Block Definition diagram that separates software and hardware; and

  d. specifications of appropriate combined fragments in relevant Sequence diagrams.
4. Subsequently, develop a SoS interaction model capturing interoperations among the set of three IRMS and the regional TCC for local traffic control through a Sequence diagram.
5. Identify and define external interfaces using the Use Case diagrams and Sequence diagrams; and then model the interfaces in the Block Definition and model their structure and behaviour in a corresponding Internal Block diagram.
6. Evaluate all diagrams in terms of their syntactical and semantic correctness, completeness against and traceability to the narratives, and consistency between model elements defined in different diagrams. Revise the model subject to the evaluation outcome.
7. Perform a study on traffic flow analysis and determine how traffic flow can be modelled mathematically and how critical flow and density can be calculated.
8. Provide the following design specifications and annotate them for the relevant model elements, which include
  a. IRMS software: high-level control strategies over the local traffic;
  b. IRMS hardware: design constraints on properties; and
  c. external interfaces: what information, energy or material is exchanged and how.
9. Capture the above specification in a SysML Requirement diagram with appropriate traceability to their source model elements.
10. Define a set of preliminary test cases to verify the specifications; and then model in the Requirement diagram with the proposed details of the test cases modelled in UML/SysML behaviour diagrams.
11. During the realisation of the above produce, capture all assumptions and justification made to support the modelling and architecting of the system.
12. Summarise all of the artefacts created into a succinct project solution and prepare a presentation for the acquiring stakeholder.

11.3.2 CLASS ASSESSMENT

- Correctness and completeness of the models. Specifically, this would mean
  - correct usage of UML/SysML syntax and semantics in all diagrams created;
  - complete and accurate modelling of alternative behaviours of the IRMS;
  - correctly ordered execution of functions and exchange of messages in the SoS interaction model;
  - correct identification and definition of external interfaces;
  - complete and accurate modelling of the properties and operation of the hardware, and their internal structure;
  - a complete and accurate set of specifications and design constraints captured in the SysML diagrams; and
  - accurate definition and modelling of a representative number of preliminary test cases.

- Sufficiency and accuracy of analyses, which covers accurate and detailed:
  - analysis of critical flow, density, and rate equations; and
  - proposal of local traffic control strategies;
- Desired qualities of the system model, which are
  - traceability of model elements to the narrative and the specifications;
  - consistency between the UML/SysML diagrams;
  - explanatory text on key points, architectural decisions, and assumptions made; and
  - application of iterative and recursive model development.
- Appropriateness of the overall approach to the case study, which includes
  - work planning, breakdown, and allocation; and
  - presenting the package to stakeholders in an appropriate format.

11.3.3   System Concept Description

At this stage of concept development, the IRMS concept does not yet have an 'intelligent' feature. To include such an intelligent feature, two additional operational modes shall be introduced:

- An adaptive mode whereby the IRMS will determine an ad-hoc control strategy in response to local traffic. This would mean a variable phasing for the traffic signal leading to a dynamic ramp admittance rate compared to the fixed-time mode (see Section 11.1.3).
- A collaborative mode whereby the IRMS deploys a dedicated strategy received from and determined by the TCC.

The IRMS shall have the ability to switch between the fixed-time mode and the adaptive mode based on its own 'decision' subject to traffic analysis by the IRMS using local traffic data. However, the TCC has the ability to override the currently deployed mode of any IRMS within its regional regime at any time. The IRMS shall resume to the fixed-time mode when the TCC withdraws the deployment of a collaborative mode to that IRMS.

    As a proof of concept and to demonstrate how different modes work in a System of Systems context, the following scenario has been defined for the project with a conceptual illustration provided in Figure 11.7. Briefly, the scenario involves a TMSoS consisting of a single TCC that has control authority over three IRMS in its region. IRMS-1 and IRMS-2 are located on two distant ramps feeding vehicles into a dual-lane motorway, respectively; while IRMS-3 is located on a ramp that is feeding vehicles into another single-lane motorway. The two motorways merge at their ends into a three-lane motorway.

11.3.4   Instructor Guidance

Before starting the tasks, the student should complete the study of the third tutorial (Chapter 7) in Part II.

    The first step in the iterative revision of the architecture is to analyse the additional narrative provided in 11.3.3 to identify the alternative behaviours of the system. Specific questions that should be answered include: When does the system need to decide on which behaviour to follow? What is different in the alternative behaviours from the basic behaviour? What is the condition for following an alternative behaviour? What happens after the alternative behaviour has completed? The answers to these questions will eventually form the basis for the new model elements to be created in the architecture.

Once the narrative is analysed and understood, the case study should move on with revising the UML diagrams following the transformational method outlined in Chapter 7, Section 7.5 and demonstrated in Chapter 7, Section 7.1. The starting point would be to capture the functionalities required to achieve the alternative behaviour(s) by extending use cases in the IRMS system-level Use Case diagram; and also, to capture at what point these functionalities would be used by extension points in the extending use cases. An example here could be a "Deploy Adaptive Mode" use case as opposed to the default "Deploy Fixed-time Mode" use case, with a possible description of the extension point as "Fixed-time Mode Inappropriate".

Followed by the revision to the Use Case diagram, the Use Case description(s) created in the first part of the Case Study can be updated accordingly by specifying the details of the alternate flow. This then directly translates into the corresponding Activity diagram(s). The challenge here is to appropriately synthesise the basic flow and the alternate flow(s) using a pair of decision and merge nodes (see Chapter 8, Section 8.3 for a detailed explanation and more examples). The modelling of the multiple paths should be consistent with the updates made in the Use Case diagram. For instance, in the example above, the specification of the extension point as "Fixed-time Mode Inappropriate" would require an action to evaluate the appropriateness of the mode deployed (likely to be based on traffic data) and evaluate its outcome to be "Mode Inappropriate" to enable the continuation to the alternate flow for deploying the adaptive mode. An (partial) Activity diagram capturing this example of mode evaluation is provided in Figure 11.8 for reference. Obviously, there could be other ways to architect the alternative behaviours especially when the second possible alternative, the collaborative mode, comes in. It is therefore important for the student to make reasonable assumptions and justify the architectural decision made.

While revising the IRMS Activity diagram to include alternate paths, the IRMS Class diagram can be revised concurrently to reflect the need of additional properties and operations of relevant system elements as modelled in classes. If necessary, additional system element(s) might be introduced. This is a point where separate modelling of hardware and software using Block Definition diagram and Class diagram respectively can be useful in complexity management. With the introduction of new system elements, the IRMS Activity diagram would be further revised concurrently to capture additional functional allocation and to be re-synthesised to ensure consistency. This step is an iteration of the procedure in the second part of the Case Study (Section 11.2).

The last step in the model revision is to update the process model as captured in the Sequence diagrams. As alternate flows are specified in the concordant Activity diagram, the Sequence diagram should elaborate the details of the alternate flows using combined fragments such as alternatives (refer to Chapter 9, Section 9.2 for more details). Other types of combined fragments, such as loop, should also be appropriately used to fully capture and explain the control structure.

Through the iteration as described so far, the IRMS models become more complete with model traceability and consistency being maintained incrementally. The next step in

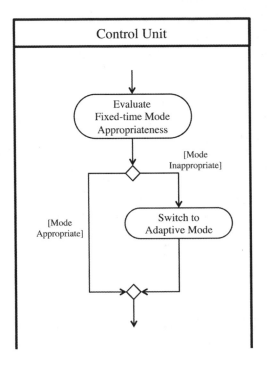

FIGURE 11.8   A example IRMS Activity diagram capturing an alternate path

the tasks is to use the single IRMS models to develop an interaction model for a system of three IRMS and one TCC. The concept of model reusability is the key in achieving this objective. Instead of recreating another set of models for the second and the third IRMS, one could simply reuse the existing IRMS models to continue the SoS modelling through the proper instantiation of the IRMS classes. Certainly, this would require making the critical assumption that the three IRMS have the same internal system behaviour and structure but could differ on how they interact with external elements. Given that the project seeks to deploy such an IRMS widely nationally, the assumption might be justified through the economic benefit brought by the standardisation of the IRMS. Figure 11.9 provides a starting point for the SoS Sequence diagram showing how the instantiation is done.

In the completion of this SoS Sequence diagram, the following questions need to be addressed:

• Is the Sequence diagram capturing the complete set of behaviours of how the three IRMS and TCC could interact? As an assurance activity, the student could test the completeness by eliciting and testing possible scenarios, e.g., the TCC requires IRMS-1 and IRMS-2 to deploy dedicated collaborative modes but not commanding IRMS-3 while IRMS-3 has locally decided to use an adaptive mode.

Following a refinement of the IRMS architecture and development of the TMSoS architecture, external interfaces can be identified and defined using the existing

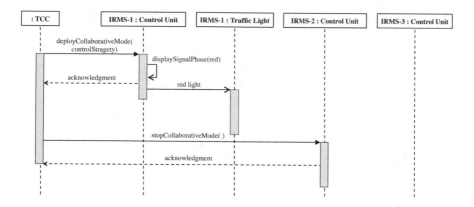

FIGURE 11.9   An initial SoS Sequence diagram as a starting point

Use Case diagrams and Sequence diagrams based on the method demonstrated in Chapter 7, Section 7.2. The definition of the interfaces should be captured as part of the SysML Block Definition and Internal Block diagrams pair developed previously. Once the interface definition and modelling are completed, summary of the interface specification can be merged into the Requirement diagram to be developed later when software and hardware specifications are also made available.

To enable system design in the next phase of the systems engineering lifecycle, architectural level analysis is a mandatory process. Following the completion of the models, the next task is to perform these analyses. Specifically, the student needs to search and study literature on mathematical modelling of traffic flows. Such mathematical models vary greatly in fidelity: some are mathematically simple and intuitive, while other are much more sophisticated and profound. The choice of the model depends on the decisions to be made. At this point, the idea is to make an initial recommendation for the design specifications on traffic control strategies rather than actually designing the strategies. This should be considered when making the choices on how traffic flow should be modelled at the architectural level.

Applying such a mathematical model to the scenario depicted in Figure 11.7 together with reasonable assumptions on initial conditions such as ranges on flow rates expected during normal operational hours and rush hours on the motorways and on individual ramps, high-level traffic control strategies can ben proposed. The strategy should clearly illustrate:

- The conditions for when an IRMS would switch to an adaptive mode and what is the range on admittance rates.
- The conditions for when the TTC would override local IRMS operation modes and what is the range on admittance rates.

The proposed architecture and control structure, together with the implementation models, form the software specification. Details of the control strategy should be properly annotated on the graphical models where relevant, e.g., mode switching conditions, and specified as a design in a SysML Requirement diagram.

Concurrently as software specification is being formulated, a hardware specification can also be proposed through recommending design choices within the derived constraints on hardware properties, e.g., constraints on the separations of inductive loops. Again, additional research should be done if needed. However, the learning objective is not to focus on accuracy of the constraints, but to understand the technical process and make reasonable recommendations subject to justifiable rationales, for instance, why the proposed limits on the separation between an inductive loop and the IRMS control unit is considered appropriate.

Following the completion of the specifications, propose a set of preliminary test cases that could be used to test these specifications when the system is fully developed. For example, how could the traffic control strategies as in the software specification be tested? Model the methods of the test cases in behavioural diagrams such as Activity diagram, and; then link the diagrams to the test case specification in the Requirement diagram. Finally, specifications presented in the Requirement diagram should utilise various relationships, such as «refine», to capture traceability between each other and also to other model elements in the system models.

With a complete set of specifications, a final stakeholder meeting can be carried out which will be followed by the initiation of the design phase or formulation of purchasing contracts if the meeting is successful. To demonstrate the readiness for design, the final project solution should, at a minimum, include all the up-to-date models; demonstration of model traceability to the original stakeholder needs as captured in the narratives; demonstration of model quality through documented model evaluation; specifications that have traceable origins to the model elements in the graphical models and architecture analyses; and last but not least, an executive-level summary similar to the one in Chapter 7, Section 7.4 to facilitate stakeholder meetings and communications.

12 Case Study
System of Actuator Systems Architecture

KEY CONCEPTS

System modelling
Model complexity management
Composite structure
Architecture trade-off

This case study provides the reader with an opportunity to further practice system architecting and design through the development and evaluation of various architectures for a System of Actuator Systems to improve reliability. There is a shift of focus in this case study to a purely mechanical system instead of an information-intensive system such as in the Chapter 11 case study and the tutorials in Part II. This shift will not diminish the role of architecture and systems in the solution of the problem but instead will emphasise their role in finding and understanding a solution to address requirements that are derived from speciality engineering, such as systems safety engineering.

In addition, a specific learning objective of this case study is for the student to develop a sense of the 'art' of architecting. As a system becomes more complex, model complexity increases. Without proper complexity management, graphical models could eventually become unreadable which would defeat the purpose of facilitating communications between stakeholders. Clever usage of modelling concepts, such as composite structure, is the key to developing an architecture that is complex in content, but simple and intuitive in presentation.

The chapter is organised into two parts each providing case study instructions similar to the previous case study. The first part concerns the modelling of a single Actuator System while the second part evolves this system into a System of Actuator Systems that will address safety requirements. The architectural solution to this problem complements the constraint example in Chapter 3, Section 3.3 and demonstrates the power of the essential definitions. Three structural types for the system of systems architecture will be investigated by the student: parallel, series, and hybrid. The property of reliability in this case study that is specified for system safety cannot be successfully implemented in all three of these types of structure. The solution is to be found in understanding which of the types can support the specified reliability property.

DOI: 10.1201/9781003213635-12

12.1   SINGLE ACTUATOR SYSTEM ARCHITECTURE

This first part of the case study models a simple Actuator System (AS) using SysML diagrams.

12.1.1   Class Exercise and Procedure

To achieve the above aim, use a professional modelling tool to produce a system model consisting of:

- A system environmental model and behaviour model capturing how the AS stabilises a load using a Use Case diagram and Activity diagram
- A system structure model capturing the relevant (system and environmental) elements and their internal structure using a Block Definition diagram and Internal Block diagram pair

To complete these tasks, the following procedure can be adopted:

1. First, understand how the system works based on the system description provided in 12.1.3.
2. Then, start the modelling of the AS by creating a system-level Use Case diagram to capture the functionalities with appropriate functional hierarchy and how they associate with the elements in the environment of the AS.
3. Subsequently, create an Activity diagram to capture the functional flow(s) in achieving the desired functionality(-ies), model the functional allocation, and concurrently define blocks in a SysML Block Definition diagram to capture the system elements and their operations.
4. Complete the SysML Block Definition diagram with the inclusion of elements of the environment, properties of these blocks, the system hierarchy, and the relationships between the system elements.
5. Using the Activity diagram as a reference, elaborate the Block Definition diagram into an Internal Block diagram to capture the connection of the system elements and their (mechanical) interfaces.
6. Evaluate the diagrams in terms of syntactic and semantic correctness, completeness against and traceability to the narratives, and consistency between model elements defined in different diagrams. Revise the model subject to the evaluation outcome.

12.1.2   Class Assessment

The system model shall be assessed by:

- Correctness and completeness. Specifically, this means
  - correct usage of SysML syntax and semantics
  - a complete set of functionalities of the AS and interactions with the environment

- an accurate and complete representation of the actuation mechanism
- a complete set of system elements for the AS, their properties and hierarchy
- an accurate representation of the internal structure
- Desired qualities of the system model, which are:
  - traceability of the model elements to the system description
  - consistency between the SysML diagrams

12.1.3 SYSTEM CONCEPT DESCRIPTION

A suspension actuator is one of the simplest suspension systems. It is widely used to stabilise a load e.g., suspension system used in vehicles to maintain a stable posture while moving through a changing road condition. Figure 12.1 is a conceptual diagram that depicts how such an actuator system works. It is important to note that in this description, the actual system behaviour is conceptualised through simplification of the dynamic and continuous nature of the actuation into the flow of displacements.

As shown, the actuator consists of three elements:

1. an Absorber;
2. a Fixed Connector with a fixed length, L_f; and
3. a Variable Connector with a varying length, L_v, in its initial state

On the one hand, the Absorber (e.g., a damper) is connected to the ground via the Fixed Connector (e.g., the axle of a wheel of a car); while on the other hand, the absorber is connected to the load via the Variable Connector (e.g., a suspension arm). When the system moves from an initial state to a compensated state where the vertical position of the ground is displaced by Δx, the actuator system maintains the vertical position of the load by:

1. Transferring the displacement through the fixed connector to the absorber.
2. Absorbing the displacement by creating a compensated displacement, Δx_c. In a fully functional situation, $\Delta x_c = -\Delta x$.

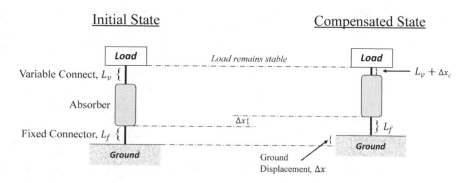

FIGURE 12.1 Conceptual illustration of how suspension actuator works

3. Applying the compensated displacement to the variable connector, such that the length of the variable connector is changed to $(L_v + \Delta x_c)$.
4. Transferring the resultant displacement to the load causing it to displace by $(\Delta x + \Delta x_c)$, where this displacement equals to zero in the fully functional situation.

12.1.4 INSTRUCTOR GUIDANCE

Essentially, the first part of this case study is about applying the same semantic transformational processes as in the previous case study but to a much simpler system that consists of only three elements in which the properties are already pre-defined in the system description in Section 12.1.3.

In the system-level Use Case diagram, it is evident that there are two elements in the environment that are interacting with the system and that the system has one purpose, which is maintaining the position of the load. Based on the description, it is also clear how this purpose is achieved through a set of steps. These steps reveal the functional decomposition and can therefore be captured in both the Use Case diagram and the Activity diagram, with proper specification of lower-level use cases and corresponding actions. One question worth consideration is how the function, *"transfer the resultant displacement"*, should be captured in both diagrams. In the scenario where the actuator functions as intended, the resultant displacement is expected to be zero. Likewise, when the actuator fails, the resultant displacement will be non-zero. Different ways of capturing this function, for example, a more generic definition of scenario without the specification of post-condition as opposed to separating possible (working/failing) scenarios, can obviously lead to a different architecture and more importantly, will have different effects on how safety analysis can be performed on the models.

Because elements of the system and their intended functions are clearly given in the description, functional allocation can be made and should be reflected properly in both the Activity diagram and the SysML Block Definition diagram. The development of the remaining parts of the Block Definition diagram is also relatively straight-forward based on the system description and the conceptual illustration in Figure 12.1. As SysML Blocks facilitate two different ways of representing hierarchical structure (see Chapter 10), it is important that the diagram is specified in an appropriate way to clearly depict *all* the information that has been provided in the system description.

Following the completion of the Block Definition diagram, the next step concerns elaborating it to reveal the internal structure of the system elements in a SysML Internal Block diagram, while maintaining the consistency of the behaviour model developed and presented in the Activity diagram.

Once the models are completed, it is necessary to evaluate the models in terms of completeness, traceability, and consistency, with respect to each other and the system description. This case study will not have a section explicitly on iterative architecture development as in the previous case study in Chapter 11.

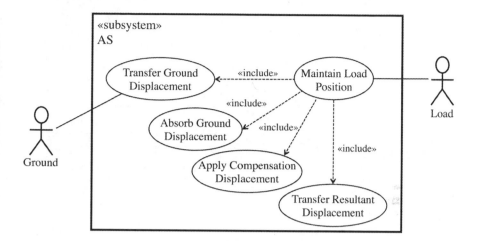

FIGURE 12.2   Sample solution: AS Use Case diagram

Therefore, alternative scenarios should be considered here, if relevant. These will refer to the previous architectural decision that the student has made on how the function, "*transfer the resultant displacement*", should be modelled.

Before moving onto the next part, the following set of diagrams (Figures 12.2–12.5) are provided as a sample solution to the modelling of the AS. The student is advised to attempt to model the AS before studying the provided answers.

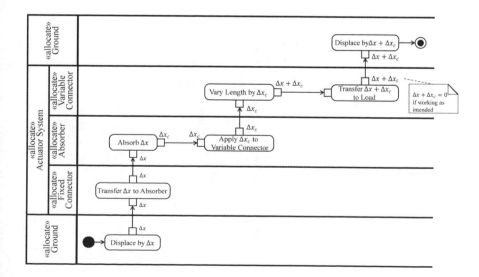

FIGURE 12.3   Sample solution: AS Activity diagram

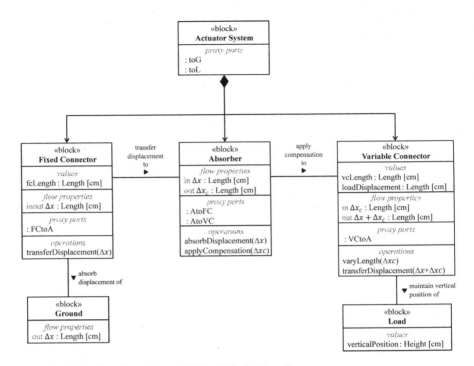

FIGURE 12.4   Sample solution: AS Block Definition diagram

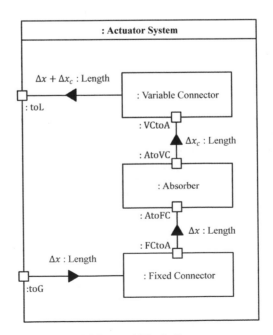

FIGURE 12.5   Sample solution: AS Internal Block diagram

12.2 MULTIPLE ACTUATOR SYSTEM ARCHITECTURE

The second part concerns with building a System of Actuator Systems (SAS) consisting of multiple actuators to address safety requirements derived from safety analyses.

12.2.1 CLASS EXERCISE AND PROCEDURE

To achieve the above aim, the following specific objectives (as tasks) are set:

- Propose two possible architectural solutions for the new SAS each consisting of two Actuator Systems. For each of the proposed architecture solutions:
  - model the system behaviour and behaviour using a set of SysML diagrams as in the previous part; and
  - demonstrate the effects of the given failure modes in the appropriate SysML diagrams.
- Summarise how the proposed solutions address the safety requirements in Table 12.1.
- Based on the findings in the above tasks, propose at least one new architecture that would satisfy the safety requirement, "*the system of actuator systems shall have no single point of failure*". Model this architecture using the concept of *composite structure* (see Chapter 10, Section 10.2) and demonstrate how it addresses requirements on both failure modes.
- Reflect on all the proposed architecture so far and make your final recommendation subject to a trade-off analysis.

To complete these tasks, the following procedure can be adopted:

1. If the work is to be completed by an individual, plan the work appropriately; otherwise, assign a Chief Architect who should further partition and allocate the work to team members in addition to planning the work.
2. Start with the safety analysis summary provided by safety engineers in 12.2.3 to understand the failure modes of the system; thereby annotate their effects in the Activity diagram and Internal Block diagram of a single Actuator System as provided in Figures 12.3 and 12.5, respectively. These

TABLE 12.1

Safety Analysis of Various Architecture

Requirement	No Single Point of Lock Failure	No Single Point of Loose Failure
Series Architecture	*Yes/No?*	*Yes/No?*
Parallel Architecture	*Yes/No?*	*Yes/No?*

should be used as the basis for the demonstration of effects of failure modes in other architecture.

3. Create a system model for a series 2-Actuators System (2-AS) architecture following the same procedure provided in 12.1.1.

4. Annotate the 2-AS Sequence diagram to demonstrate how each failure mode propagates through the exchanges of messages (displacements) with a reference to Step 2 above.

5. Repeat Steps 3 and 4 for the parallel 2-Actuators System architecture.

6. Summarise your findings in Table 12.1.

7. Propose and model a SAS consists of multiple actuators (at least 2 but no more than 4) that would successfully address both failure modes and demonstrate how the failure modes are addressed as in Step 2, 4, or 5.

8. Perform an architectural trade-off analysis by comparing all the architectures that have been proposed, i.e., single, series, parallel, and multiple actuators, with respect to standard and expected design objectives such as safety, reliability, cost, performance etc., for the vehicle in which the AS shall be used.

12.2.2 CLASS ASSESSMENT

The package shall be assessed through the following four viewpoints:

- Correctness and completeness of the architecture. Specifically, this would mean
  - correct usage of SysML syntax and semantics in the SysML diagrams created;
  - accurate modelling of a series 2-AS architecture capturing a complete system behaviour and structure;
  - accurate modelling of a parallel 2-AS architecture capturing a complete system behaviour and structure;
  - accurate modelling of the n-AS architecture using composite structures and successfully addresses the safety requirement; and
  - correct and sufficiently detailed demonstration of effects of the failure modes for each proposed architecture.
- Desired qualities of the system model, which include
  - consistency between the SysML diagrams.
- Architectural analysis, which includes:
  - accurate summary of the requirements satisfaction as captured in Table 12.1; and
  - well-justified architectural trade-off analysis.
- Transferable skills, which include
  - applying a standard modelling language to a range of systems problems in an end to end system architecture problem to develop a project technical package; and
  - work planning, breakdown, and allocation; and
  - presenting and justifying the final solution to stakeholders in an appropriate format.

i) Lock Failure   ii) Loose Failure

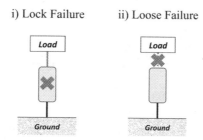

FIGURE 12.6   Actuator system failure modes

12.2.3   SAFETY ANALYSIS SUMMARY

A system-level safety requirement has been defined which states that *the actuator system shall have no single point of failure*. A safety specialist, upon investigating the architecture of the AS proposed in the previous part, has determined there are two failure modes that are considered as a single point of failure: lock failure and loose failure. As depicted in Figure 12.6, the two failure modes are explained as follows:

   i. A *lock failure* is when the absorber element is considered 'locked', i.e., the displacement cannot be absorbed. As such, the compensation does not take place. The ground displacement is passed through the absorber and applied to the variable connector. Effectively, this is the same as having $\Delta x_c = 0$. This could lead to undesired events in which the severity depends on the magnitude of the ground displacement and its effect on the load.

   ii. A *loose failure* is when the variable connector is broken. As such, not only can Δx_c not be applied, but the load also has the potential to hit the absorber and or other consequences that lead to a catastrophic system-level failure.

Therefore, to avoid a single point of failure, the safety specialist has recommended using a redundant architecture, i.e., *stabilise the load using more than one actuator*.

12.2.4   INSTRUCTOR GUIDANCE

For the first task, having understood the failure modes in terms of their associated components and consequences, it is then possible to annotate the model elements and explain how the failure modes contribute to system-level failures. To help the reader organise their thoughts, consider the following questions. First, which component is associated with the lock (and loose) failure? The answer to this question will lead to an annotation to the corresponding part in the Internal Block diagram. Then, which operation of that component will fail as an immediate effect? The answer to this question will lead to an annotation to the corresponding action in the Activity diagram. Finally, how does this effect propagate, thereby leading to the final failure consequence? The answer to this question will lead to demonstration of the effect of the failure mode using the functional flow in the Activity diagram.

   After completing this model-based failure mode analysis, the next set of tasks is concerned with developing models for two different 2-AS architecture and performing the same failure mode analysis to these architecture to demonstrate how well they address the failure modes.

Taking the series 2-AS architecture as an example, the reader should consider the following questions when developing the architecture using the single AS model as the basis:

- Will the Use Case diagram change?
- How the Activity diagram and Block Definition diagram should be changed? Make sure that the consistency between these diagrams is maintained.
- Finally, according to how the Block Definition diagram is changed, make appropriate changes to the Internal Block diagram to maintain consistency.

For the modelling of the 2-AS, the student reader is invited to attempt to apply the concept of composite structure from here onward but could also wait until developing the n-AS architecture.

The annotation of failure modes would follow the same spirit as for the single AS. It is important to note that for completeness, failure modes could happen to either of the two actuators in the 2-AS. The final finding should be captured in Table 12.1. The correct answer should show that neither of the two types of 2-AS will successfully address both failure modes. Therefore, the n-AS architecture is required.

The reader can be creative in terms of proposing the n-AS architecture given that $2 < n \leq 4$. However, based on the analysis result captured in Table 12.1, it should be clear what would be an appropriate proposal. The modelling and safety analysis to follow should justify this. If the analysis invalidates the architecture proposal, the reader should continue with proposing and analysing other options repeatedly until a valid solution is found.

In the demonstration of how the failure modes are addressed by the n-AS, if composite structure was not applied, the reader would find that the demonstration will be difficult as the model complexity grows quickly with the number of actuators growing. Therefore, the modelling task for the n-AS should focus on investigating how composite structure can be applied. Specifically, the idea is to avoid explicit modelling of each actuator and its components by unique blocks, but to reuse the single-AS model and use hierarchical instantiations. For instance, this would mean the creation of two Internal Block diagrams, one at the n-actuator level and the other at the level of a single actuator. Since the actuators are identical in terms of structure and behaviour, the actuators at the system level will trace to the same internal structure. However, an important revision is necessary here. The provided Internal Block diagram in Figure 12.5 is specifically for a single-AS that is 'connecting' to the ground on one hand and the load on the other; this is unlikely to be the case for an n-AS system. For the Activity diagram, the concept of composite structure as well as model reuse can be applied.

Although there are n-AS solutions that successfully address the failure modes, is this really a preferred solution? The final objective then is to investigate this further by performing a trade-off study among the architecture solutions that have been proposed so far. In addition to safety objectives, other design objectives should be reasonably considered. The final solution should be well justified with a sound argument and ideally with supporting evidence.

Part IV Bibliography

Dickerson, Charles E., Siyuan Ji, and Rosmira Roslan. 2016. "Formal methods for a system of systems analysis framework applied to traffic management." In *2016 11th System of Systems Engineering Conference (SoSE)*, (June 2016): 1–6. IEEE.

Ingram, Claire, Zoe Andrews, Richard Payne, and Nico Plat. 2014. "SysML fault modelling in a traffic management system of systems." In *2014 9th International Conference on System of Systems Engineering (SOSE)*, (June 2014): 124–129. IEEE.

Annex A-1: Essential Mathematics

KEY CONCEPTS

Axiomatic set theory
Predicate calculus
Tarski model theory
Category theory

The mathematical basis of the essential definitions of structure, architecture, and system introduced in Chapter 3 and used throughout the book are established in the axiomatic theory of sets and classes in mathematics. The object-oriented Unified Modeling Language (UML) standardised by the Object Management Group also traces back to class theory but has been adapted to software development. Chapter 2 introduced the Tarski model theory and its basis in set theory and the first-order predicate calculus of mathematical logic. This annex provides a succinct summary of the mathematics of these theories without burdening the reader with detailed mathematical proofs. These are covered in Sections A-1.1 and A-1.2, and are sufficient for the first-order model theory that underlies science and engineering.

Architecture though is concerned with structures and properties of, or assigned to structures. As explained at the end of Chapter 4, the concept of architecture used in the essential definitions is a relation between properties which doesn't quite fit into the constructs of first-order logic. The deeper understanding of architecture in terms of structure is based on Category Theory. This was addressed by Eugenie Hunsicker (Dickerson et al. 2019) and will be covered in Section A-1.3.

A-1.1   CLASS AND SET THEORY

The essential mathematics needed to reason logically about the definitions and concepts of systems and architecture is the axiomatic theory of classes and sets. Every axiomatic theory is organised around undefined terms whose meaning are established by statements called axioms. Intuitively, a class $C(p)$ consists of all objects having the property p. The objects of the class are also referred to as members of the class.

In a mathematical theory of sets and classes, there are two undefined terms whose meanings are determined by the axioms of set theory. These are *class* and the *dyadic relation* '$\in$' of membership between two classes. In the Bernays–Gödel–von Neumann axiomatic set theory, there are eleven axioms. This is just one axiomatic theory (the other most notable being Zermelo–Frankel). It provides a logically consistent theory of sets. A succinct summary for the practitioner with a minimal background in mathematics can be found in Dugundji (1966).

In set theory, the term *property* has a very narrow meaning that is associated with class definition. The formal relation between property, class, and the dyadic relation (i.e., *element of*) is the formula $p(A): A \in B$, where A and B are classes. Further expressions are built up from such formulae by negation, conjunction, disjunction, and quantification of the class variables (for example, $\exists A: p(A)$) by means of the predicate calculus. The Axiom of Class Formation ensures that every class of sets is uniquely determined by its defining property $p(x)$, where x is a set. The mathematical objects ensured by the axiom are then sets.

Sets are specialisations of classes. Specifically, only those classes that can be members of a class are defined to be sets. This technical construct is used in order to avoid antinomies that can arise in Class Theory, such as the Russell paradox. Heuristically, a class can be thought of as any collection of objects specified by some property; and a set can be thought of as a class that can be regarded as single entity (an abstraction that admits logical or physical existence). Technically, a set can refer to physical objects but in itself, it is not a physical object. A set is an abstraction.

Sets are, therefore, normally discussed in the context of a predefined Universal Set, which is denoted as U. The process of defining a set S follows the Axiom of Class Formation: if p is a property such that each element of U either does or (exclusively) does not have the property p, then all of the elements x belonging to U that have the property p form a set, which is denoted by $S = \{x \in U: p(x)\}$. The term property is associated with a declarative statement $p(x)$; as in the Propositional Calculus, e.g., 'x has the property p'. The members (objects) of a set are referred to as *elements* of the set.

It is important to note that despite the one-to-one pairing of a class (or set) with a property, the collection and property are not one and the same. This distinction is important in engineering, for example, where a *set of cars* might be attributed the property of being *fast*.

It can be important in engineering (including object-oriented software development) to regard all members of a set that have a property which specifies a unique characteristic to be a single entity, i.e., to be in some sense equivalent. When this is done, the term *property* is specialised to be the term *type*. In object-oriented software standards this is common practice, e.g., data types; and the term type is usually associated with the terms *characteristics* and *features* of the objects that are intended to give the objects an identity. In natural language, this may seem to be 'shades of grey', but in mathematics, equivalence can be given a precise meaning (binary relations that are reflexive, symmetric, and transitive). This use of the term type should not be confused with other uses such as Type Theory that specify constructs of syntax to avoid the antinomies that can arise in Class Theory, or to provide hierarchical classifications in software. However, it is appropriate to think of structural type as a classifier in the (class) category of structures. Note that 'relational' is a structural type and within this type of structure are familiar mathematical structures such as algebraic and topological structures.

A-1.2 LOGIC AND MODEL THEORY

Modern mathematics considers Model Theory as the study of classes of mathematical structures from the perspective of mathematical logic. Model, class, and theory (hypothesis) are three of five elements of scientific explanation, the other two being entity and observation. Theory is expressed as system of sentences in a formal language (Overton 2012).

In the 1930s, there was an opinion that semantic notions such as truth and denotation should not be incorporated into a scientific conception of the world. Tarski's work (1983) on truth challenged this opinion by introducing formal models for representing the semantics of logic. Specifically, Tarski extended the interpretation of formulae in the propositional calculus of logic, in which Hartry Field (1972) argued that what Tarski succeeded in doing was to reduce the notion of truth to certain other semantic notations.

The Propositional Calculus as a language of mathematical logic is concerned with the calculation of truth. (The Calculus of Newton and Leibniz, on the other hand, refers to the calculation of limits.) It formalises that part of logic where the validity of an argument depends only on how sentences are composed and not on the internal meaning of the propositions. For this reason, it is also called the Sentence Calculus. It provides the basic syntax and rules for constructing well-formed formulae that mirror the Boolean algebra of set theory. However, it lacks the syntax and semantics necessary to reason about relations; and therefore, is inadequate for a theory of models.

The syntax of the Predicate Calculus follows that of the Propositional Calculus, although it includes new symbols such as predicate letters to represent relations, equality ($=$), and quantifiers. Specifically, the Universal Quantifier over a variable x is denoted as $\forall x$, which is read as 'for every x', and the Existential Quantifier over a variable x is denoted as $\exists x$, which is read as 'there exists x'. A *sentence* in the Predicate Calculus is defined to be a fully quantified well-formed formula. The syntax and rules for the construction of sentences in the Propositional Calculus is followed in the Predicate Calculus by, for example, replacing propositional variables with predicate letters.

As noted in the main body of the text, Tarski model theory offers the following simple but formal definition: a model is a relational structure for which the interpretation of a sentence in the Predicate Calculus becomes valid (true). In a purely logical and set-theoretic expression the validity of the model derives from the relational structure being a non-empty set. The simplicity of this formalism though is blurred when predicates, and their variables and constants are given semantics from natural language or from domain ontology. The term *semantic predicate* has been used in this book to describe such situations. This is the case in conceptual structures and graphs. The criteria of the relational structure being non-empty require further interpretation of the terms.

For example, in the statement 'there exists a fast car', which is a fully quantified predicate in natural language; several more details need to be provided before a truth value can be rendered. The set of cars in the discourse needs to be specified along with key characteristics, e.g., acceleration and top speed. A criterion for fast also needs to be specified, i.e., does fast only mean acceleration? Measurable values must also be specified.

There are also fundamental limitations of the first-order predicate calculus, which need to be considered and adhered to. For systems and conceptual structures, the most significant is that this calculus is based on relations between variables (as well as constants) but not between relations. The formula $P(Q(v))$ is not a well-formed *first-order* predicate. This is not to say that compositions of functions are not used routinely in mathematics and engineering. Compositions of this type are formulated in terms of variables rather than relations of relations. The use of mathematical compositions and their relation to structures will be addressed in the next part of this annex.

To summarise, the idea of interpretation is central to the validation of theories (sentences) in mathematics, and has been extended in this book to the validation of definitions, e.g., by interpreting definitions of *architecture* and *system* into mathematical structures. Tarski model theory is sufficient to provide a foundation for reasoning about the essential definitions offered in this book, but it should be expected that a semantically richer theory will be needed for a more complete architectural theory of engineering systems. This is to be expected, given the development of concepts in mathematics and systems theory over the twentieth century. At the beginning of the century, as the concepts of set theory began to be formalised, it was common for a set of elements to be regarded as a system. Dedekind, for example, regarded sets in this way in his well-known treatise on number theory (Dedekind 1901). At the end of the century, Lin, for example, proposed that a set M endowed with a collection of relations R to be a (general) system (Lin 1999). The members of M were referred to as the objects of the system. His definition though, $S = (M, R)$, in first-order model theory is only a relational structure. The Bertalanffy notion of a system (Bertalanffy 1969) is similar to the Lin 'general system' but is semantically richer. The importance of structure to both definitions is clear. In the second half of the twentieth century, notions of structure further evolved into Category Theory.

A-1.3   CATEGORY THEORY

Category Theory was first introduced in the 1940s by mathematicians S. Eilenberg and S. MacLane (1945) as a tool in algebraic topology. The original conception of this subject in its birthplace of mathematics is as a tool for studying classes of structures in algebra and topology and the relationships among different structural views of the same objects through maps between categories, called functors. Thus, it is very natural to introduce Category Theory into a discussion of architecture, as categories are a way in which to define architecture within pure mathematics, where categories can be thought of as architectural types. The objects of a category are members of a class (e.g., of spaces) and junction between classes is accomplished by the functors. This is consistent with the position of Rosen (1993), who proposed Category Theory as a framework in which the process of modelling can be captured.

In recent years, Category Theory has also been used for constructing models in various disciplines. In computer science, it has been used for modelling in programming language semantics and domain theory (Pierce 1991); and in systems for modelling networks such as electrical circuits and chemical reaction networks (Fong 2016). It has been proposed by D. Luzeaux (2015) as a formal foundation for systems modelling. While this work is interesting, constructing categories as models misses the original potential of Category Theory as a framework for a much more fundamental study of systems and their architecture.

Category Theory relates to both class and set theory and to logic and Model Theory but enriches these in terms of the study of structural properties and relations. By definition, a category is an ordered pair (O, M), where O is a collection of objects and M the union of collections of maps (or morphisms), called $\text{Hom}(A, B)$ for any two objects A, B which are members of O, that satisfy certain rules of composition. Note importantly that the collection of objects, O, is not required to be a set. In fact, an important category SET, the category of sets, does not form a set (i.e., as noted in A-1.1, the 'set of all sets' is a class that is not definable as a set).

Category Theory expands on Class Theory in an important way of relevance to architecture: it includes structural information about the collection through the specification of maps between the objects, morphisms that preserve their essential structure. For example, in the category $\text{Vect}(R)$, whose objects are real vector spaces, $\text{Hom}(A, B)$ is the set of all linear transformations between A and B. This is a category; it conforms to the structural type Category.

Category Theory also relates to, but in an important way, expands on Model Theory. In Model Theory, a model is the image of a theory in a relational structure defined on a set. However, the essential definition of architecture requires the additional possibility of specifying relations on the relations, which are called *interrelations*. These arise both in Use Case diagrams on their own, and, for example, in describing the interrelation between a Use Case diagram and an Activity diagram. Thus, Model Theory is not an adequate mathematical theory for expressing Architecture Definition. Relational structures do form a category, and this can be refined to form a new category of interrelational structures: a (second level) interrelational structure is a set A together with a set R_1 of relational structures on A and a set R_2 of relational structures on R_1. This can be continued to permit arbitrarily many levels of interrelations. For notational simplicity, write R_0 for A.

In order to tie interrelational structures back to Model Theory, it is useful to use a slightly different description of a model in logic than usual, as a triple (A, σ, I_A) rather than as a pair (A, R). The set A is the same in the two descriptions. The set σ is a collection of symbols representing potential relations, each with a specified order (unary, binary, ternary, etc.). The set sigma is called the *signature* of A. I_A is called the *interpretation map* for A, which assigns each element of σ to a relation on A. The image of σ under I_A in the first description then forms the set R of relations in the second description. A set of logical sentences involving the symbols of σ is called a *theory* in the signature σ. The interpretation map I_A can be extended to a map from any theory in σ to an isomorphic set of sentences related to the relational structure, $R = I_A(\sigma)$, on the set, A. The triple (A, σ, I_A) is a model of the theory if the isomorphic sentences are true. Note that this relates to Sowa's work on ontology: a theory in the predicate calculus starts out as an ontologically neutral string of symbols in the language of predicate calculus enriched by the symbols of σ. It is given an ontology through the interpretation map that relates this language to the domain of the relational structure.

Now an interrelational structure $(A = R_0, R_1, ..., R_n)$, gives a sequence of relational structures (R_{j-1}, R_j). Each of these can be rewritten as a triple, (R_{j-1}, σ_j, I_j). Now a sequence of logical theories $(T_1, T_2, ..., T_n)$, in the signatures $(\sigma_1, \sigma_2, .., \sigma_n)$ can be specified. An interrelational structure will be a model of a vector of theories if each triple (R_{j-1}, σ_j, I_j) is a model of T_j in the sense above. In other words, making the isomorphisms required by Model Theory explicit, it can be extended to collections of theories relevant to the pairs in an interrelational structure; and thereby create the richer set of structures and models needed for a mathematically based Architecture Definition process.

Architecture Definition can then be understood as follows in a mathematically rigorous way:

- Define a sequence of theories $\{T_j\}_{j=1...n}$ relevant to an interrelational structure $(A = R_0, R_1, ..., R_n)$, described in natural or technical language (or both), and relevant to a Conceptual Graph that formalises a concept.
- Specify a sequence of signatures $\{\sigma_j\}_{j=1...n}$ for the language(s) associated with the theories.
- Specify a fully interpreted first order model using the interpretation of the theories $\{T_j\}_{j=1...n}$ in terms of the relational structures $\{(R_{j-1}, R_j)\}_{j=1...n}$

Note that here the collection of interrelational structures of the fully interpreted models of the theories for a particular concept will form the objects of a category associated with the concept, where the morphisms are those maps between objects which preserve the signature of the theories.

Note that the process of architecting can begin with theories, structures, or signatures and progress to either of the two others. It is possible to start with a concept then realise it in a vector of logical theories, and construct referents that model the vector of theories. This would describe the process of 'green-field' architecting, in which properties are first specified and objects satisfying them are sought. However, it is also possible to start with a category of fully interpreted models, where the collection of objects O is simply defined by its members and there are some ideas of what parts of one object map to what parts of another (morphisms), but the defining properties of the collection may not yet be understood or articulated. This would correspond to reverse architecting, in which an object or collection of objects is specified and the fundamental underlying properties of them are sought, or may arise in 'brown-field' architecting, in which the process starts with some theories and models and is required to make these concordant.

Finally, there is an important difference between Category Theory and both class (and set) theory and model theory. Category Theory has the concept of functors. A functor is a map between categories in which both objects and morphisms are transformed from one type to another. As mentioned in the first paragraph, functors are a way of changing viewpoints by mapping objects of one structural type to objects of a different structural type. This makes Category Theory particularly appropriate as a mathematical language in which to formulate and study a formal theory of architecture and systems, which involves transformations of model types through both changes of viewpoint and model refinement.

In order to formalise a theory of architecture and systems, it is necessary to define categories with more information, such as measurable properties associated to elements or to relations or interrelations. This structural type would define a category of measurable interrelational structures **M-Int-Struct**. It is also possible to assign a set of states to given elements, as well as a description of how the element will move through these states over time. This will involve incorporating graphs associated to components or relations into the category, giving **State-M-Int-Struct**. Finally, it is necessary to discuss a process whereby the structures in these categories are refined as the architecting process is carried out. Thus, the overall category will have a graded structure corresponding to level of detail: **State-M-Int-Struct(n)**. In terms of these categories with various attributes, various architectural transformations can be described, as between architectural views, or as between architectures described at different levels of detail.

BIBLIOGRAPHY

Von Bertalanffy, Ludwig. 1969. *General Systems Theory.* New York: George Braziller Inc.

Dedekind, Richard. 1901. *Essays on the Theory of Numbers.* Chicago: Beman, WW Open Court.

Dickerson, Charles E., Michael Wilkinson, Eugenie Hunsicker, Siyuan Ji, Mole Li, Yves Bernard, Graham Bleakley, and Peter Denno. 2019. "Architecture Definition in Complex System Design Using Model Theory." arXiv:1909.06809v3.

Dugundji, James. 1996. *Topology.* Boston: Allyn and Bacon, Inc.

Eilenberg, Samuel, and Saunders MacLane. 1945. "General theory of natural equivalences." *Transactions of the American Mathematical Society* 58, no. 2: 231–294.

Field, Hartry. 1972. "Tarski's theory of truth." *The Journal of Philosophy* 69, no. 13: 347–375.

Fong, Brendan. 2016. "The Algebra of Open and Interconnected Systems." Doctoral thesis, University of Oxford.

Lin, Yi. 1999. *General Systems Theory: A Mathematical Approach.* New York: Kluwer Academic/Plenum Publishers.

Luzeaux, Dominique. 2015. "A formal foundation of systems engineering." In *Complex Systems Design & Management,* edited by Frédéric Boulanger, Daniel Krob, Gérard Morel and Jean-Claude Roussel, 133–148. Cham: Springer.

Overton, James A. 2012. "Explanation in Science," Doctoral thesis, The University of Western Ontario.

Pierce, Benjamin C. 1991. *Basic Category Theory for Computer Scientists.* Cambridge : MIT Press.

Rosen, Robert. 1993. "On models and modeling." *Applied Mathematics and Computation* 56, no. 2-3: 359–372.

Tarski, Alfred. 1983. *Logic, Semantics, Metamathematics: Papers from 1923 to 1938.* Indianapolis: Hackett Publishing.

Annex A-2: Logical Modelling of Language

KEY CONCEPTS

Natural language
Predicate calculus sentences
Interpretation of language
Relationships and properties

This annex details the procedure that was used in Chapters 3 and 4 to produce logical models of the terms system and architecture. As noted in the chapters, this is analogous to how sentences in first-order models are interpreted into relational structures but without explicit reference to the formal language of the predicate calculus. The procedure was originally presented by Dickerson (2008) and then again in the first edition of the book (Dickerson and Mavris 2010). Several examples were provided in these two sources that demonstrate how to interpret concepts expressed in a natural language definition into the graphical structure of a Unified Modeling Language (UML) diagram. Similar examples of this type of interpretation of concepts date back two decades in both international standards and commercial practice, e.g., in the IEEE Standard 1471 (IEEE 2000) and the commercial practice of architecture (Hatley, Hruschka and Pirbhai 2000); and continue into present times.

This annex will provide the details of how the definition of system represented in Figure 3.1 of Chapter 3 was derived. The level of detail using predicates in the chapter exceeds any previous explanation of the formal understanding of using UML diagrams as logical models. To synthesise predicates in an intuitive way into abstract Class diagrams, minor modifications to the original procedure have been made.

A-2.1 PROCEDURE FOR THE MODELLING APPROACH

The objective of the logical modelling of a sentence is to extract the relations comprising the sentence to derive a minimal model that is complete and captures its intended meaning. Key words will be selected that remain undefined to avoid the circularities possible in natural language. This approach follows the axiomatic approach of mathematics. The structure into which the relations are interpreted will determine the degree of formality of the model of the sentence. A structure that is mathematically defined can result in a model that is more formal, whereas graphical representations of the structure, such as UML diagrams, are less formal but can be more intuitive.

When deriving a logical model of a sentence, it is important to avoid introducing information that does not belong to the intended meaning of the sentence. The model should capture the exact meaning of the sentence within the limitations of natural language and the intent of the sentence.

The modelling of a sentence proceeds in three steps. The first is to list the key words that express its meaning, then organise the words around nouns and verbs, and finally to interpret the organisation of the words into the selected structure or diagram. This procedure can be applied to the core structures represented graphically in the notations of UML. This is how verb-noun phrases were used in Chapter 3 to express use cases as predicates.

- The key words will be listed and remain undefined; their meaning will be determined by relations between the words. They can be 'listed' using a mark-up of the sentence.
- In general, nouns are represented as classes (of objects) and verbs are represented as relations. This is similar to how predicates were used in Chapter 3 to explain the diagram in Figure 3.1.
- An adjective generally expresses a property of an object. This is best represented as a noun which is a property of the objects of a class.
- The primary type of diagram used from UML is the abstract class diagram which provides notations for representing classes and relations.

The approach and notation described by Dickerson (2008) leveraged the notations of UML but with certain conventions to track the key words, changes, and rationale. The notations are not standard in UML but can be useful for marking up an abstract class diagram to annotate deviations from the exact statement of the concept being modelled. Specifically:

- Natural language notation
  - The **concept (defined term)** uses:
    - Bold font
    - Underlined
    - Not italics
  - The *key words* in the concept use:
    - Bold font
    - Not underlined
    - Italics
    - Capital for nouns
  - *Other words* in the concept use:
    - Not bold font
    - Not underlined
    - Italics
  - Additional words (i.e., not specifically occurring in the sentence) use:
    - Not bold
    - Not underlined
    - Not italics

- Graphical notation
  - Nouns are placed in boxes (classes) and capitalised.
  - Verbs and relations are placed on lines.
  - Solid boxes and lines are used for key words.
  - Dash-dot graphics are used otherwise.

Note that whilst a class is a collection, the name of a class is by convention always a singular noun. Adjectives are converted to properties (nouns possessed other nouns) or relations (e.g., verbs) provided the intended meaning is preserved, unless their role gives logical scope that would be violated by the conversion. Depending on the readability of the text, bold font might not be used for discussion of the sentence and key words in the natural language discourse of a diagram.

The procedure is finished when all of the key words (both cited from the sentence and necessary for the completion of the model) have been identified and all of the relations between the words (both explicit and intended by the natural language) have been captured in the diagram.

The criteria for the quality of a logical model include the intrinsic consistency and completeness of the model, and the correctness of the model relative to the natural language under accepted interpretations (e.g., into concrete examples). It is also desirable to have independence between the relationships and independence between the key words; unless this violates the intended meaning of the sentence.

A-2.2 DEMONSTRATION OF THE PROCEDURE FOR MODELLING THE TERM *SYSTEM*

The sentence that was used to define the term system in Chapter 3 was taken from ISO/IEC/IEEE 15288: 2015 (ISO 2015). Specifically, *a system is*

> *a combination of interacting elements organised to achieve one or more stated purposes.*

Based on this definition, the term system is seen to be comprised of seven key words: *combination, interacting, elements, organised, achieve, stated,* and *purpose*. As noted in Chapter 3, the adjective *stated* can be suppressed in the initial stage of modelling. To interpret it as a verb calls for a predicate subject such as stakeholders (who state the purpose); to interpret it as a property of *purpose* is awkward in natural language. The procedure does provide a notation for allowing the introduction of a term not specifically occurring in the sentence (such as stakeholders); but for the sake of simplicity and minimalism, this will not be pursued in the demonstration. The procedure then results in a mark-up of the sentence: *a* **system** *is*

> *a* **combination** *of* **interacting** **elements** **organised** *to* **achieve** *one or more stated* **purpose***(s).*

Note that *stated* has not been marked up and will need to be dealt with separately.

The second step in developing the logical model would be to capture the direct relationships starting with those between the defined term *system* and the other key words, and then capture the other relationships. The list of six predicates in Chapter 3 has already done this. The two directly related to system are:

- System achieves Purpose
- System is a Combination

The next obvious relationship in the sentence is:

- Combination of Elements

This accounts for four of the six key words that have been identified for the model. The other two are adjectives that will be converted into nouns that become properties: organisation and interaction. The phrase 'interacting elements' then becomes,

- Elements have Interactions

The property of organisation subsequently applies to both the interactions and the combination:

- Combination has Organisation
- Interactions have Organisation

The six terms and six relationships are captured in the UML class diagram in Figure A-2.1. This is precisely the diagram in Figure 3.1 but with the notations that are pre-scribed by the procedure.

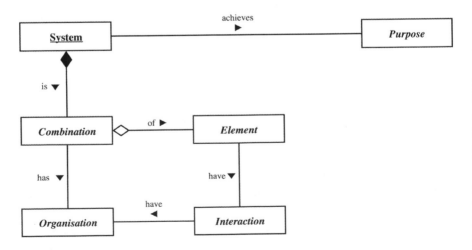

FIGURE A-2.1   Logical model of the term system

BIBLIOGRAPHY

Dickerson, Charles E. 2008. "Towards a logical and scientific foundation for system concepts, principles, and terminology." In *2008 IEEE International Conference on System of Systems Engineering*, (June 2008): 1–6. IEEE.

Dickerson, Charles E. and Dimitri N. Mavris. 2010. *Architecture and Principles of Systems Engineering*. New York: CRC Press.

Hatley, Derek, Peter Hruschka, and Imtiaz Pirbhai. 2000. *Process for Systems Architecture and Requirements Engineering*. New York: Dorset House Publishing.

IEEE (Institute of Electrical and Electronics Engineers). 2000. *IEEE recommended practice for architectural description for software intensive systems*. IEEE Std 1471-2000.

ISO (International Organization for Standardization). 2015. *System and software engineering – System life cycle process*. ISO/IEC/IEEE 15288:2015.

Annex A-3: Terminology for Architecture and Systems

KEY CONCEPTS

Systems engineering standards
System structuring
Action and interaction
Context and environment

Chapter 3 introduced foundational terminology for architecture and systems that was key to the practice of engineering. As noted in Chapter 3, it is important to understand that the standardisation of concepts, terminology, and processes is an agreement between users, practitioners, and other stakeholders of the standard. The current *de facto* standard for systems and software engineering is ISO/IEC/IEEE 15288:2015 (ISO 2015). The INCOSE *Systems Engineering Handbook* (INCOSE 2015) was published in parallel with the publication of ISO/IEC/IEEE 15288:2015. It seeks to elaborate the standard for the purpose of industrial practice and provides an extensive commentary. These two coupled sources offered definitions for only two of the six key terms in Chapter 3, system and system architecture. Essential definitions of these two terms were offered based on summative research by Dickerson et al. (2021) together with complementary definitions of structure, model, functionality, and engineering. This annex provides a lexicon for more than a dozen other terms that are routinely used in practice. These terms support the technical processes for system and architecture definition.

A-3.1 RELEVANT TECHNICAL PROCESSES

Historically, systems engineering has employed *system definition and decomposition* to define system elements and architecture. The current ISO/IEC/IEEE standard refers to system decomposition only twice and does not offer a definition. However, in its definition of system, it does refer to a "combination of interacting elements". Clearly, whenever a product or service is regarded as a combination of elements, something has in fact been *decomposed* into elements. Thus, the concept of system definition and decomposition is a natural concept based on the definition of system that has been adopted in the standard.

Decomposition defines a system hierarchy and is therefore part of any architecture definition process. This can be blended with the system requirements definition processes to produce a technical view of the system that implements the concept of definition and decomposition. As presented in Chapter 4, the salient features of each of the two processes are as follows.

Requirements Definition employs the following concepts to specify mission and stakeholder needs in a technical view of the system.

- System boundary, elements, functionality, and processes
- Functional, performance, and non-functional requirements; and constraints
- Traceability of system requirements to stakeholder requirements

Architecture Definition is concerned with both stakeholder and system requirements, and system structure. It also enables Design Definition. The salient features are:

- How structure and properties enable functionality
- Definition of external interfaces; identification of element interfaces

A-3.2   A FURTHER LEXICON OF TERMS

Definitions will now be offered for the following key terms that have been used in the tutorial, the essential technical processes, and the structured methods:

Abstraction
System Boundary
Definition and Decomposition
System Hierarchy
Interaction
System Behaviour
Context
Environment (operational and operating)
Operational Concept
Operational Requirement
Scenario
System of Systems

Definitions of each of these terms follow below in the order above, which provides a better logical flow than alphabetical order. The terms in the left column are generally associated with concepts of structuring a system.

A-3.2.1   ABSTRACTION

Abstraction is the suppression of irrelevant detail. It is a function of the human mind and is the key to modelling and the reduction of complexity in conceptualisation. As put forth by Victor Frankl (2006) in the third school of psychology in Vienna, the mind constantly seeks to make sense (seek meaning) of perceptions. Abstraction relates to meaning and can be accomplished by

- Generalisation
- Deletion

- Distillation
- Distortion

These methods of abstraction may be complete, but they are not independent. Note that language itself is an abstraction.

A-3.2.2 SYSTEM BOUNDARY

In the *Systems Engineering Handbook* (INCOSE 2015, 6), the definition is:

> *The system boundary is a demarcation between a system and its environment that defines which elements in the context belong to the system and which do not.*

This definition aligns well with its usage in ISO/IEC/IEEE 15288: 2015, but the standard does not offer a definition. However (in paragraph 6.4.3.3), the standard does consider a functional boundary of the system that is used to establish expected system behaviour. Various concepts of boundary relate to system interfaces.

A-3.2.3 DEFINITION AND DECOMPOSITION

Historically, systems engineering has been based on a repeated application of a process of *system definition and decomposition*. However, neither ISO/IEC/IEEE 15288:2015 nor the INCOSE *Systems Engineering Handbook* offers a definition. Nonetheless, whenever a product or service is regarded as a combination of (interacting) elements, it has in fact been decomposed into elements. Partitions of a set are also an example. Decomposition defines a system hierarchy and is part of the architecture definition process. The definition and decomposition process can also be applied to services, system processes, and functionality.

A-3.2.4 SYSTEM HIERARCHY

The term hierarchy is broadly used in different contexts. In business or military organisations, it can refer to levels of authority such as chain of command. There is no definition in the relevant standards but the *Systems Engineering Handbook* (INCOSE 2015, 7) summarises the concept for systems as follows:

> *Hierarchy within a system is an organisational representation of the system structure using a partitioning relation.*

A-3.2.5 INTERACTION

Interaction is a key term in many definitions of terminology for architecture and systems. Although it is used extensively in ISO/IEC/IEEE 15288:2015 and the INCOSE *Systems Engineering Handbook*, it is not defined in either. The term derives from 'action', which means 'something that is done'. This is a fundamental concept in engineered systems because they 'do something'. This type of verb-noun phrase (i.e., do something) can be used to model system actions and functionality. These are the building blocks of models represented in UML Activity and Use Case diagrams, for example, in the first tutorial (Chapter 5).

Dictionary definition considers an interaction to be an effect that two things have on each other. In science and medicine, it is a specialised term that is consistent with this idea. For usage in engineering, this concept needs a more precise definition. The following definition is proposed:

> *Interaction is a type of action in which two or more entities influence or have an effect upon one another.*

Note that 'interaction' is a type of relation between entities. The term entity is used here in the sense of something that exists (Dickerson 2013); and expresses inclusiveness of systems, persons, organisations, etc. Note that actions, in general, need not involve two entities.

A-3.2.6   SYSTEM BEHAVIOUR

In a dictionary definition, behaviour is defined as the way in which one acts. The INCOSE *Systems Engineering Handbook* suggests two compatible concepts based on the following:

* Systems have both attributes and processes
* System states are values of the attributes and (binary) status of the processes (i.e., idle or active)
* System states have variations (because the system does something)
* Systems have interactions in relation to the environment under specified conditions

These concepts can be integrated as follows:

> *System behaviour comprises change in system state (attributes and processes) and the effects that its environment has on it due to interactions.*

ISO/IEC/IEEE 15288:2015 offers no definition of behaviour.

A-3.2.7   CONTEXT

The term *context* has a variety of definitions and related meanings that go beyond its usage for engineering. For example, in written or verbal communication, it can be something that clarifies or explains a statement. For usage in systems engineering, the following definition is offered that leverages the more general usage:

> *Context comprises the circumstances that form a setting of an idea, statement, object, or event in terms of which it can be more fully understood.*

ISO/IEC/IEEE 15288:2015 does not offer a definition but this definition aligns well with its usage in the standard.

A-3.2.8   ENVIRONMENT

In Section 4.1.19, ISO/IEC/IEEE 15288:2015 offers the following definition:

The environment of a system is the context determining the setting and circumstances of all influences upon a system.

There are variants not defined in the standard:

- The *operational environment* of a system is that part of the environment with which the system has interactions.
- The *operating environment* is that part of the environment with which the system interoperates to achieve it intended purpose.

These variants generally understood in this way align well with their usage in the standard.

A-3.2.9   OPERATIONAL CONCEPT

In Section 4.1.25, ISO/IEC/IEEE 15288:2015 offers the following definition:

An operational concept is a verbal and graphic statement of an organisation's assumptions or intent in regard to operation of or a series of operations of a system or a related set of systems.

The *statement* gives an overall picture from the perspective of the users and operators. This term should be not confused with the term concept of operations.

A-3.2.10   OPERATIONAL REQUIREMENT

This is a widely used term in requirements definition and analysis but neither ISO/IEC/IEEE 15288 nor the INCOSE *Systems Engineering Handbook* offers a definition. However, it can be aligned to the System Requirements Definition technical process which has been summarised as, "transformation of stakeholder and <u>user</u> views into a technical view of the system ... *that achieves its intended use in its intended operational environment*". Given that systems have behaviours, a reasonable definition would be,

An operational requirement specifies an operation or behaviour that a system must perform or have to achieve its intended use in its intended operational environment.

This is in good agreement with the way that Use Case and Activity diagrams are used as well as the definition of operational environment.

A-3.2.11   SCENARIO

The term scenario is another widely used term for which neither ISO/IEC/IEEE 15288 nor the INCOSE *Systems Engineering Handbook* offers a definition. In software testing, scenarios are hypothetical stories that help the tester systematically work through a number of related individual test cases.

The Handbook (INCOSE 2015, 195) associates scenarios (and scenario analysis) with mission use cases to analyse stakeholder needs. This is closely related to the concept of an *operational scenario*. These are typically provided in the early stages of the development process by the user or the sponsor and later refined for validation testing. They provide a coarse description of the intended situation and its dynamics. This type of scenario can evolve over time as circumstances or requirements change. The key elements are the initial state, the desired end state, the course of actions to reach the prescribed end state, and the associated entities with their capabilities and relations.

Two other types of scenarios important to system development are *conceptual* and *simulation scenarios*. These can bridge the gap between the early stages of the development process and the later stage of validation testing. Simulation scenarios require much greater detail than is provided in operational concepts. Conceptual scenarios specify the details of real-world situations that can bridge operational and simulation scenarios.

A-3.2.12 SYSTEM OF SYSTEMS

The Annex of ISO/IEC/IEEE 15288:2015 offers the following succinct definition:

> *A system of systems (SoS) is a system of interest whose elements are themselves (independent) systems (which are referred to as constituent systems).*

The term (independent) has been added to emphasise that the constituent systems of the SoS are not subsystems. An (*ad hoc*) SoS is one formed to accomplish a task that constituent systems cannot do on their own. There are different types of SoS: virtual, collaborative, acknowledged, and directed. See also the INCOSE *Systems Engineering Handbook*.

BIBLIOGRAPHY

Dickerson, Charles E. 2013. "A relational oriented approach to system of systems assessment of alternatives for data link interoperability." *IEEE Systems Journal* 7, no. 4: 549–560.

Dickerson, Charles E, Michael Wilkinson, Eugenie Hunsicker, Siyuan Ji, Mole Li, Yves Bernard, Graham Bleakley, and Peter Denno. 2021. "Architecture definition in complex system design using model theory." *IEEE Systems Journal* 15, no. 2: 1847–1860. doi: 10.1109/JSYST.2020.2975073.

Frankl Viktor E. 2006. *Man's Search for Meaning*. Boston: Beacon Press.

INCOSE (International Council on Systems Engineering). 2015. *System Engineering Handbook*. 4th ed. New Jersey: John Wiley & Sons.

ISO (International Organization for Standardization). 2015. *System and software engineering – System life cycle process*. ISO/IEC/IEEE 15288:2015.

Annex A-4: Research Impact and Commercialisation

KEY CONCEPTS

Complex systems
Relational frameworks
Digital features
Commercial utility

This annex offers a brief review of the past decade of research that underlies the book and explains how the processes and methods have been developed from government-sponsored research and academic collaboration with commercial industry. As mentioned in Chapter 1, systems engineering originated from defence and aerospace engineering where projects were mainly concerned with complex systems that had long lead times for development. The decade of research that underlies the book has been a leading effort in the United Kingdom for taking a critical step towards a successful advancement of systems engineering from its origin in defence to the commercial world. Sectors such as the automotive industry, where systems continue to grow more complex, have benefited from and can expect a continued benefit from the adoption of systems engineering and advanced concepts for the architecture and design of complex systems.

The annex proceeds in three chronologically ordered parts. The first is a short summary of the academic research sponsored by the Royal Academy of Engineering from 2007 to 2012. This led to a substantial collaborative research programme with Jaguar Land Rover (JLR) from 2013 to 2018 that was co-sponsored by the Engineering and Physical Sciences Research Council (EPSRC) in the UK. A summary of the accomplishments was published in a joint article with EPSRC and JLR in *Impact* (Dickerson and Ji 2018). Highlights from this article form Section A-4.2. The annex concludes with current work on the commercialisation of wireless charging of electric vehicles sponsored by the government Office for Zero Emissions Vehicles (OZEV) through Innovate UK.

A-4.1 RAEng CHAIR OF SYSTEMS ENGINEERING

The Royal Academy of Engineering (RAEng) in the UK in collaboration with BAE Systems and EPSRC sponsored a Chair of Systems Engineering at Loughborough University during 2007–2012. The stated objective of the research programme was to establish a scientific basis for systems engineering and viable methods by the end of the appointment and then be employed in a new programme of application to industry. Thus, two milestones were established for the long-term objectives of research and commercial application. An early proof of concept for such a scientific basis was demonstrated (Dickerson 2008).

The focus quickly became the development, formalisation, and application of the Relational Oriented Systems Engineering Technology Trade-off and Analysis (ROSETTA) framework, which was developed in collaboration with Professor Dimitri Mavris at the Georgia Institute of Technology, US. This was complemented by ongoing work with the Object Management Group (OMG) and the International Council on Systems Engineering (INCOSE) to advance the state-of-the-art in the then-burgeoning field of model-based systems engineering (MBSE). The first edition of this book (Dickerson and Mavris 2010) captured the initial research progress but was focused on applications to aerospace and defence. This also marked the beginning of a research-informed programme of postgraduate studies in architecture and system engineering.

Extensive engagements with the international systems engineering community during the early development of MBSE led to a summative paper that pointed to relational orientation as a preferred approach to future development (Dickerson and Mavris 2013). This work was complemented by an early demonstration of ROSETTA in a substantial case study on data link interoperability (Dickerson 2013).

The first milestone was, therefore, accomplished by the end of the five years of research sponsored by the Academy (2007–2012) but many details remained to properly formalise the mathematical basis of ROSETTA. Nonetheless, this advanced systems engineering framework and the early methods were at a level of maturity suitable for engineering application to the challenges faced in the automotive and aerospace industries. Thus, the gateway to the second milestone for the long-term research objective for commercial application was opened by the Programme for Simulation Innovation (PSi) from 2013 to 2018 that was co-sponsored by JLR and the EPSRC.

A-4.2   THE EPSRC/JLR PROGRAMME FOR SIMULATION INNOVATION

PSi was a five-year research programme funded by JLR and the EPSRC. It was a £10 million multi-university collaborative programme originally organised around eight research themes across several UK universities over the period of 2013 to 2018. The aim was to develop a capability for virtual prototyping through simulation to provide manufacturers in the UK access to new, world-class simulation tools and processes. Loughborough University led Theme 1, the "Analysis of the Vehicle as a Complex System".

The advanced framework and methods developed under the RAEng Chair were well-matched to this theme of the PSi research programme. The ROSETTA framework is a model-based formalism that lends itself to the graphical languages of UML and SysML (the Unified and Systems Modeling Languages). The formalism of the framework, rooted in the Tarski theory of models in mathematical logic, had been adapted to the practice of engineering for model specification and relational transformation for the purpose of system description, analysis, and design. This provided a significant advancement over the traditional hierarchical decomposition methods currently practiced by systems engineers.

The rise of the digital economy and the associated increase in demand for customised products had caused the modern premium automotive vehicle to become a complex system. Integration has increasing levels of influence on innovation for manufacturing processes. The research on the complexity of digital features concentrated on advanced integration facilities for virtual integration and verification. The objective was "Get it right the first time through good design for speed to market". As the digital enablement of vehicles becomes more complex, automotive manufacturing requires new innovative approaches for modelling and simulation with improved analysis tools in order to successfully integrate existing technologies and processes alongside new technologies to meet increasing market demand. Each one requires successful integration within the vehicle as a complex system of systems. Thus, the strategic intent of the research for the PSi was to use innovations in modelling and simulation to evolve state-of-the-art capabilities in vehicle design and analysis for integration of the digital features of a high-interaction multi-disciplinary complex vehicle.

Two case studies were defined: (i) engine emissions reductions and (ii) vehicle motion management. Both case studies addressed a challenge that faces automotive manufacturers: the problem of meeting customer expectations in the context of design constraints. The case study on engine emissions reduction addressed the customer expectation for lower fuel consumption while manufacturers must adhere to emissions regulations that are becoming increasingly strict. The case study on vehicle motion management looked at the longstanding research field of vehicle motion control. The outputs of this case study were coordinated control architectures that overlay onto a distributed electronic architecture.

The research on coordinated control architectures was published in the work by Lin et al. (2018). An integration of UML with the propositional calculus of logic for the reliability of safety-critical systems was published in the work by Dickerson, Roslan and Ji (2018). The application of ROSETTA to the engine emissions reduction case study resulted in a patent for a calibration system and method (Dickerson, Ji and Battersby 2018). An international standard for the UML Profile of ROSETTA (UPR) was adopted in 2018 and finalised in the following year (OMG 2019).

The academic Principal Investigator for "Analysis of the Vehicle as a Complex System" was Professor C.E. Dickerson. Dr David Mulvaney, Senior Lecturer in embedded microelectronic systems, was the academic Co-Investigator. The industrial Principal Investigator was David Battersby, the Senior Manager for Software Architecture at JLR. Dr Siyuan Ji was the lead academic Research Associate and is now Lecturer in safety critical systems engineering at the University of York.

A-4.3   IUK AMICC PROJECT FOR WIRELESS CHARGING OF ELECTRIC VEHICLES

The advances made in the EPSRC/JLR PSi programme became the basis for an exploitation path to apply architecture trade-offs and optimisation to an innovative commercialisation project on electric vehicle wireless charging led by a consortium of small to medium enterprise companies and university partners sponsored by the government office OZEV through Innovate UK. The two-year demonstration project (2020–2022) is called AMICC.

The main objective of AMICC is to evaluate the benefits of static wireless charging systems through the deployment of an innovative charging system to enable fleet vehicles with short dwell times and high utilisation to switch to electric vehicles when they would otherwise have been unable to do so. This is supported by the development of an innovative hardware solution to provide standardisation across the industry for high-power smart charging.

The Loughborough University role includes modelling the architecture of the demonstration systems and the design space in a way that can be used to explore the design envelope beyond the demonstration system subject to the constraints of standards, regulations, and science for acceptance of evolving technologies for future integration into the wireless charging system concept.

BIBLIOGRAPHY

Dickerson, Charles E. 2008. "Towards a logical and scientific foundation for system concepts, principles, and terminology." In *2008 IEEE International Conference on System of Systems Engineering*, (June 2008): 1–6. IEEE.

———. 2013. "A relational oriented approach to system of systems assessment of alternatives for data link interoperability." *IEEE Systems Journal* 7, no. 4: 549–560.

Dickerson, Charles E. and Dimitri N. Mavris. 2010. *Architecture and Principles of Systems Engineering*. New York: CRC Press.

———. 2013. "A brief history of models and model based systems engineering and the case for relational orientation." *IEEE Systems Journal* 7, no. 4: 581–592.

Dickerson, Charles E. and Siyuan Ji. 2018. "Analysis of the vehicle as a complex system, EPSRC." *Impact* 2018, no. 1: 42–44.

Dickerson, Charles E., Siyuan Ji and David Battersby. 2018. Calibration system and method. U.K. Patent GB2555617, filled November 4, 2016, and issued May 09, 2018.

Dickerson, Charles E., Rosmira Roslan, and Siyuan Ji. 2018. "A formal transformation method for automated fault tree generation from a UML activity model." *IEEE Transactions on Reliability* 67, no. 3: 1219–1236

Lin, Tzu-Chi, Siyuan Ji, Charles E. Dickerson, and David Battersby. 2018. "Coordinated control architecture for motion management in ADAS systems." *IEEE/CAA Journal of Automatica Sinica* 5, no. 2: 432–444.

OMG (Object Management Group). 2019. *UML Profile for ROSETTA (UPR)*, Version 1.0.

Annex A-5: Using This Book for a One-Semester Module of Lectures

This annex provides a reference for educators who plan to use this book as the basis to develop and deliver a module in model-based systems engineering with a focus on System and Architecture Definition. The content presented in this annex is based on more than a decade of teaching experience. By United Kingdom (UK) Higher Education (HE) standards, a module like this would amount to a one-semester module in a Master-level taught programme, for example, an MSc in Systems Engineering. Nonetheless, it is also appropriate for such a module to be tailored for a Continuing Professional Development (CPD) module or a Professional Refresher Course module that targets Systems Engineers and Speciality Engineers, e.g., Safety Engineers, whose primary work influence and are influenced by Systems Engineering activities.

The annex is organised into three parts, with the first part outlining the key pedagogical approach and learning objectives, the second part providing the recommended module structure and delivery style, and the last part providing a guideline on module assessment. The material in the book has been successfully delivered to both remote and in-person classes. In either style of delivery, the book provides a 'teacher's assistant' that is more attractive to students than recorded lectures.

A-5.1 PEDAGOGICAL APPROACH AND LEARNING OBJECTIVES

A-5.1.1 PEDAGOGICAL APPROACH

The authors recommend educators adopt a practice-based learning approach, where traditional lecturing is kept at a minimum but sufficient level. Practice-based, in this context, means that students are expected to learn from hands-on experiences in modelling systems with software tools and developing system architecture in a collaborative working environment. To further enhance such a learning experience, the practical sessions (see Section A-5.2 for details) should be organised to imitate a typical workplace environment. Specifically, such a hypothetical environment refers to a group of systems engineers/architects working together to develop a system architecture against requirements from the acquiring stakeholder, which is a role to be 'played' by the lecturer. Following the completion of the practical sessions, students are invited to a 'stakeholder meeting' to present their architectural solutions to the acquiring stakeholders and to receive immediate feedback. By setting up the sessions this way, students will develop transferable skills in addition to the intended learning objectives as outlined in the subsequent section. These skills include but are not limited to: teamwork, leadership, presentation, and most critically, communication,

TABLE A-5.1

Module Structure and Mapping

Monday	**Lectures**	
	Session 1: *Introduction*	(Chapter 1)
	Session 2: *Logical and Scientific Approach*	(Chapter 2)
	Session 3: *Concepts, Standards and Terminology*	(Chapter 3)
	Session 4: *Structured Methods*	(Chapter 4)
Tuesday	**Tutorial**	
	Tutorial: *ATCS System Definition*	(Chapter 5)
	Laboratory Sessions	
	Modelling Language: *UML - Part I*	(Chapter 8)
	Practical: *TMSoS System Definition*	(Chapter 11.1)
Wednesday	**Stakeholder Meeting**	
	Student Presentation: *TMSoS System Specification*	
	Tutorial	
	Tutorial: *ATCS Architecture Definition*	(Chapter 6)
	Laboratory Sessions	
	Modelling Language: *UML - Part II*	(Chapter 9)
	Practical: *TMSoS Architecture Definition*	(Chapter 11.2)
Thursday	**Stakeholder Meeting**	
	Student Presentation: *TMSoS Architecture Specification*	
	Tutorial	
	Tutorial: *ATCS Architecture Refinement and Analysis*	(Chapter 7)
	Laboratory Sessions	
	Modelling Language: *SysML*	(Chapter 10)
	Practical: *TMSoS Architecture Refinement and Analysis*	(Chapter 11.3)
Friday	**Stakeholder Meeting**	
	Student Presentation: *TMSoS Final Solution*	
	Lecture	
	Session 5: *Conclusion*	(Chapter 4.5)

which is an indispensable competency required by system architects due to their roles and responsibilities in the systems engineering lifecycle.

Additionally, to further support the students in achieving the desired learning outcomes, teaching resources such as teaching assistants to frequently engage with the students are desirable. Timely feedback during the laboratory sessions and stakeholder meetings (see Table A-5.1) from the lecturer and the teaching assistants are considered a key contributor to students' success.

A-5.1.2 INTENDED LEARNING OBJECTIVES

The following intended learning objectives are defined based on the book. These objectives are ordered with an increasing level of specificity according to the Bloom Taxonomy of educational objectives. These are to:

1. Understand core terminologies defined for and concepts associated with model-based systems engineering.

2. Use standardised modelling languages to model systems of interest with given system information.
3. Analyse system behaviours based on developed/given system architecture.
4. Evaluate system models against desired qualities such as completeness, traceability, and consistency.
5. Develop system architecture using transformational methods and creative thinking.

A-5.2   MODULE STRUCTURE AND DELIVERY STYLE

A one-semester module at the Master-level or a typical CPD module involving students from industry should be ideally delivered as a one-week block taught module as opposed to weekly sessions spanning the entire one semester. This is to allow continuity in the learning experience to support the practice-based learning approach described in the previous part. As such, the module structure is organised around a one-week schedule and a standard proposal of such a structure is illustrated in Table A-5.1. Essentially, the module consists of four major elements: subject matter lectures, tutorial lectures, laboratory sessions that consist of modelling language training (or review) and practical case study sessions, and stakeholder meetings in the form of (student) group presentations.

Subject Matter Lectures – These lectures should cover materials in Part I, i.e., from Chapter 1 to Chapter 4. Since these chapters are foundational to the rest of the book, the material therefore should be delivered prior to other elements and become a point of reference throughout the rest of the week. In addition, the final conclusion lecture (Session 5 in Table A-5.1) draws on Chapter 4 again (specifically 4.5) to conclude the module. Each chapter among the four is appropriate to be delivered as a standalone lecture. The delivery style is preferably lecture with the lecturer presenting material using slide-based presentations.

Tutorial Lectures – The tutorial lectures should be based on the tutorial case studies in Part II, i.e., from Chapter 5 to 7. Depending on the preferred domain of application, the educator may wish to adapt the materials for a different system than the Air Traffic Control System, for example, a Railway Signalling and Control System. Despite which system is to be studied, these tutorial lectures provide a detailed end-to-end study of a sufficiently complex but intuitively understandable system. The delivery style is a standard lecture with in-class exercises (see exercises throughout Chapter 5 to 7 for details) to enable opportunities for engagement and formative test of understanding.

Laboratory Sessions (Modelling Languages) – All of the laboratory sessions should be conducted in a computer lab equipped with a licenced professional modelling tool. For the modelling language sessions, these should cover materials in Part III, i.e., from Chapters 8 to 10. The delivery style for these sessions is in the form of individual self-study, i.e., studying and practicing the modelling languages using a computer, with support from a teaching assistant(s). Learning progress is primarily driven by the completion of the exercises provided in the corresponding chapters.

Laboratory (Practical) Sessions – The practical labs are based on the Case Study materials presented in Chapter 11, with each practical session corresponding to a part in Chapter 11. Similar to the tutorial lectures, the system to be studied, which in this case is a Traffic Management System, can be modified, but should follow the same structure as in the chapter to ensure alignments with the tutorial lectures and the modelling languages sessions. Students are organised into groups to allow the development of transferrable skills such as teamwork and leadership. Each group has a designated 'Chief Architect', i.e., the group lead that is responsible for managing the 'project'. Students will work against the specific objectives following the recommended procedure as outlined in the corresponding parts of the chapter (e.g., Chapter 11, Section 11.1.1). The lecturer acts as the 'acquiring stakeholder' to clarify concerns on the system concept description. The lecturer also intervenes minimally on students' progress to maximise teamwork experience and to improve understanding and proficiency through tackling challenges. The Instructor Guidance should be used by the students wisely to enable progress. It is preferable to have teaching assistant(s) in these sessions to help students on technical problems related to the usage of the modelling tools.

Group Presentation – These sessions are for students to present their work to the 'acquiring stakeholder' (the lecturer) and their peers (other students) following the completion of a practical session. The presentation should be kept at roughly 10 to 15 minutes for each group with 5 minutes allocated to questions coming from the acquiring stakeholder and peers. The content of the presentation will be used for assessment (see Section A-5.3 for details).

A-5.3   ASSESSMENT GUIDELINE

The assessment of the module should be formed of three parts: individual exercises, group presentations, and module coursework.

A-5.3.1   INDIVIDUAL EXERCISES

Throughout the book, exercises are introduced in certain chapters for student readers to practice modelling and architecting concepts. These exercises, especially for the ones provided in the modelling languages chapters (Chapters 8–10), can be self-assessed with model answers provided by the lecturer. Alternatively, where teaching resources are available, the modelling exercises can be part of the formative assessment, where models created by the student would be assessed by a teaching assistant immediately upon completion so that timely feedback can be provided to the student.

Further to the exercises, it is also recommended that the case study presented in Chapter 12 is used as an individual or group formative assessment post the delivery of the block taught week. Compared to the case study presented in Chapter 11, this one is light weight and is considered suitable as additional but possibly optional study material. Again, depending on teaching resource availability, the assessment could be either self-assessed against provided model answers and followed by a Q&A session, or assessed by the lecturer or the teaching assistant with formative feedback provided in a timely manner to help student in completing the module course work.

A-5.3.2 GROUP PRESENTATION

For the student presentations, this can either be formative assessment or summative assessment or a mixture of both, e.g., with the first two presentations assessed formatively while having the last presentation contribute to the summative assessment. Each presentation could be assessed based on the evaluation criterion listed in the Class Assessment subparts in Chapter 11. There is no fixed marking scheme, however, an example one is provided as follows.

For each of the points in the evaluation criterion, e.g., "traceability of the UML diagrams to narrative", a Fail/Pass/Distinction scale can be utilised. For the 'Fail' grade, the group would fail to achieve the associated learning objective. In this example, this would mean that the UML diagrams do not exhibit traceability to the given narrative showing a lack of understanding of what model traceability is and how it is maintained. For a 'Pass' grade, the group would achieve the minimum requirement of the learning objective. In this example, this would mean that the students demonstrate a reasonably acceptable understanding of traceability but with UML diagrams containing elements that are not traceable to the given narrative yet without explaining why this is the case, e.g., as new and justified design commitments. For a 'Distinction' grade, the group would demonstrate an achievement beyond the associated learning objective. In this example, this would mean that all of the model elements in the UML diagrams are either traceable to the narrative or explained and justified if not. In addition, traceability is explicitly demonstrated, for example, by using a traceability matrix. Then, the final 'mark' of a presentation is cumulatively calculated based on the number of 'Fail', 'Pass', and 'Distinction', where a 'Fail' would yield a zero point, a 'Pass' would yield half point and a 'Distinction' would yield one point. This way, the final sum would align with the typical HE grading system in the UK in a way where 50% (e.g., all evaluation criterion marked as 'Pass') and above being Pass, while below 50% being Fail.

In the case where all of the presentations are counted as summative assessments, the last one should have a higher weight given the amount of work involved and the level of difficulty. Therefore, the lecturer may wish to adjust the weighting of the evaluation criterion accordingly to fairly reflect the workload.

A-5.3.3 MODULE COURSEWORK

The module coursework is the core element of the summative assessment of the module. It is an individual coursework where the student performs an end-to-end study of a particular engineering project of their own interest to develop a system architecture solution and present it in a formal written report. In a way, the content of the report would be similar to what has been developed in the practical sessions and presented in the group presentations, but with elaborations on details and appropriate documentation. To ensure that the project defined by the student is sufficiently complex and well-scoped, it is essential that the student writes a one-paragraph proposal describing the project in terms of engineering opportunity, context, and key requirements. The proposal can be presented to the lecturer for examination of suitability. Where teaching resources are available, it would be useful for the student to elaborate the agreed proposal into a proper project narrative and present it to the lecturer for final confirmation before starting the development process. This way, the risk of working on an unsuitable project can be further minimised. Essentially, this allows the student to imagine the lecturer as the 'acquiring stakeholder' again for the project that the student wishes to work on for the module coursework.

The marking scheme of the module coursework is largely an extension of the marking criterions for the group presentations, but further includes essential evaluation criterions that are typical to scientific and engineering reports. For instance, by typical content headers, marks should be attributed to:

- Introduction and Background where the engineering problem needs to be described and the opportunity explained and justified subject to research findings
- Technical Sections including but not limited to System Description, System Architecture, Architecture Analysis, Domain Analysis, and Specifications, where all the technical contents are developed, presented, and discussed
- Discussion and Conclusion where the engineering solution is summarised and critically discussed

A detailed marking scheme is not provided at this instance due to the vast difference seen in the practices across the Universities. In general, for UK Universities, a typical Fail/Pass/Merit/Distinction is regarded as appropriate for the grading of the overall quality of the report.

Index

Printed in the United States
by Baker & Taylor Publisher Services